Recent Trends in Instrumentation and Control

(RTIC-2019)

Recent Trends in Instrumentation and Control

(RTIC-2019)

Edited by

C. Shanthi

M. Mythily

ALLIED PUBLISHERS PVT. LTD.

New Delhi • Mumbai • Kolkata • Chennai • Bangalore • Hyderabad

ALLIED PUBLISHERS PRIVATE LIMITED

1/13-14 Asaf Ali Road, **New Delhi**–110002
Ph.: 011-23239001 • E-mail: delhi.books@alliedpublishers.com

17 Chittaranjan Avenue, **Kolkata**–700072
Ph.: 033-22129618 • E-mail: cal.books@alliedpublishers.com

15 J.N. Heredia Marg, Ballard Estate, **Mumbai**–400001
Ph.: 022-42126969 • E-mail: mumbai.books@alliedpublishers.com

No. 25/10, Commander-in-Chief Road, Ethiraj Lane (Next to Post Office)
Egmore, **Chennai**–600008
Ph.: 044-28223938 • E-mail: chennai.books@alliedpublishers.com

P.B. No. 9932, No. 15, 3rd Floor (Next to Vijaya Bank), 5th Cross, Gandhinagar, Karnataka, **Bangalore**–560009, Ph.: 080-41530285 / 22386239
• E-mail: bngl.journals@alliedpublishers.com / apsabng@airtelmail.in

Sri Jayalakshmi Nilayam, No. 3-4-510, 3rd Floor (Above More Super Market)
Barkatpura, **Hyderabad**–500027
Ph.: 040-27551811, 040-27551812 • E-mail: hyd.books@alliedpublishers.com

Website: www.alliedpublishers.com

ISBN: 978-93-89934-19-9

ommittee Members

ON

, Vice Chancellor, Anna University

strar, Anna University

rajan, Dean (MIT)

IRPERSON

r. N. Pappa, HoD (I.E.)

ORGANIZING SECRETARY

Dr. C. Shanthi, IE, MIT

Dr. M. Mythily, IE, MIT

NATIONAL ORGANIZING COMMITTEE

Dr. S. Renganathan, Prof. (Retd.), MIT

Dr. P. Kanagasabapathy, Prof. (Retd.), MIT

Dr. M. Chidambaram, Prof. (Retd.) IIT (M)

Dr. V. Jagadeesh Kumar, Prof. IIT (M)

Dr. R.C. Panda, Scientist, CLRI

Dr. Babji Srinivasan, Asst. Prof. IIT (GN)

Dr. S. Senthilkumar, IIT (G)

Dr. S. Ramakrishnan, Prof. IIT (M)

Dr. S. Sundaramoorthy, Prof. PEC

Dr. T.K. Radhakrishnan, Prof. NIT (T)

Dr. B. UmaMaheswari, Prof. CEG (AU)

Dr. Ramalingam, Prof. CEG (AU)

Dr. S. Baskar, Prof. TEC

Dr. K. Srinivasan, NIT (T)

Mr. Jayaharan, ISA South India Section

LOCAL ORGANIZING COMMITTEE

Dr. J. Prakash, IE, MIT
Dr. V. Natarajan, IE, MIT
Dr. D. Manamalli, IE, MIT
Dr. S. Srinivasan, IE, MIT
Dr. S. Kumar, IE, MIT
Dr. K. Latha, IE, MIT
Dr. D. Vasanthi, IE, MIT
Dr. Sabitha Ramakrishnan, IE, MIT
Dr. S. Sutha, IE, MIT
Dr. K. Kamalanand, IE, MIT
Dr. D. Kalpana, IE MIT
Dr. M. Vijay Karthik, IE, MIT
Dr. S. Meyyappan, IE, MIT
Dr. N. Vinoth, IE, MIT
Dr. A. Ganesh Ram, IE, MIT

Acknowledgement

At the outset, we express our sincere thanks to our esteemed Vice Chancellor Dr. M.K. Surappa for providing permission to conduct this national conference on 'Recent trends in Instrumentation and Control (RTIC-2019)'. We wish to express our sincere thanks to our Registrar Dr. J. Kumar for approving this program and providing all support. We extend our hearty thanks to our Dean Dr. T. Thyagarajan, for his support, advice and for felicitating the event. We would like to express our gratitude to our beloved HOD, Dr. N. Pappa, without whose support, this event would not have been materialised. We wish to thank all the faculty members of IE department whose dedication has shaped this event to perfection. Heartful thanks to all our supporting staffs for their help in the smooth conduct of this program. A special thanks to all the research scholors, PG students and UG students for actively participated in this program.

Organizing Secretary
RTIC-2019

Preface

National Conference on 'Recent Trends in Instrumentation and Control (RTIC-2019) was held in Department of Instrumentation Engineering, MIT Campus, Anna University during March 4–5, 2019. The conference is intended to discuss the state-of the-art new research results and perspectives of future developments relevant to Control and Instrumentation with participants both from academia and from industry. Its success is reflected in the papers received, with participants coming from several places in Tamil Nadu, allowing a real exchange of experiences and ideas. Every submission has been peer reviewed by 2–3 reviewers in the right area, and more than 50 submissions have been received with only 16 of them were accepted for final presentation. This conference can only succeed as a team effort, so the editors want to thank the technical committee members and the reviewers for their excellent work in reviewing the papers as well as their invaluable input and advice. Thanks also goes to the Anna University officials and Head of the Department, Instrumentation Engineering for their various support to RTIC 2019.

Organizing Secretary
RTIC-2019

Preface

[illegible]

Contents

Study on Modelling and Simulation of Clutch Actuation and Dynamics

Vidhi Shah[1], **N. Sivakumaran**[2] **and Ritesh Goyal**[3]
[1]Department of Chemical Engineering, NIT, Tiruchirappalli, India
[2]Department of Instrumentation and Control Engineering, NIT, Tiruchirappalli, India
[3]KPIT Technologies Ltd., Bangalore, India
E-mail: [1]vidhi94@gmail.com; [2]nsk@nitt.edu; [3]Ritesh.Goyal@kpit.com

ABSTRACT: *Automatic transmission is widely used in the automobile industry due to the higher comfort and lesser hassle which it offers. It is now gaining popularity in the present Indian automobile market which is fueled by the changing trend of people choosing comfort over economy. However, as demand increases, the need for better efficiency has become an important aspect for the automobile industry, which is trying to enhance the efficiency while maintaining the ease and comfort for the driver. The clutch plays an important role in both manual transmission and automatic transmission system of vehicles. Thus, enhancement of clutch dynamics plays a vital role in bridging the gap between the comfort and efficiency of automatic transmission systems. Advance modelling techniques help us understand the behavior of existing components and enable us to evaluate various method to enhance the performance. The modelling of the clutch actuation system requires obtaining equations from the hydraulic and mechanical subsystems of the actuation system which uses a pressure reducing valve. This paper depicts the equation based modelling of the clutch actuation system, which has been implemented in Simulink. Further a Simscape model of combined clutch and clutch actuation system is created to compare the results with individual Simulink model.*

Keywords: Clutch Modelling, Clutch Actuation System, Simulink Model.

1. INTRODUCTION

Automatic transmission system requires multiple clutches with the planetary gear set in order to change gears. Though dry clutches are normally used for manual transmission systems, multi plate wet clutches are preferred for automatic transmission systems. The wet clutch requires a actuation system which has been understood from[4] whereas the torque transferred for wet clutch has been derived after studying the physical momentum balance equations, effect of number of plates and the changes in coefficient of friction. Amidst all these, there has been observed a lack of a consolidated model of clutch actuation system and wet clutch. This shortcoming has been dealt with in this paper.

The modelling tools explored are Simulink and Simscape. While the former has been used to model the hydraulic clutch actuation system (see section 2.1) using an equation based approach, the latter has been used in creating a consolidated model of complete clutch with the actuation and the dynamics, using a component based approach.

The section 2 gives detailed modelling concept of the Clutch actuation system. Section 3 is for results and analysis related to simulation of the model created in the relevant tools. Lastly section 4 is the conclusion.

2. MODELLING

2.1 Clutch Actuation System

The hydraulic circuit diagram of the clutch actuation system has been shown below in Figure 1 to understand the path line of the flow and the topology of the connections.[4] The pump pumps the fluid which is highly

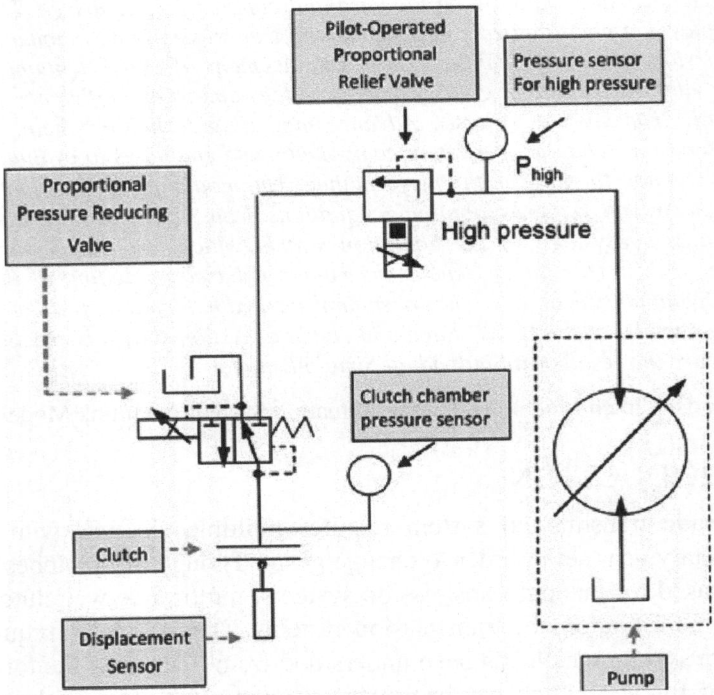

Fig. 1: Hydraulic Circuit Diagram of the Clutch-Actuation System

pressurized and forms the high pressure reference. Linked next is a proportional pressure-reducing valve, which provides the chamber pressure. The valve used to control the clutch chamber pressure is a three-way two-position proportional pressure-reducing valve. Figure 2(a) shows its cross section. Port 1 is connected to the clutch chamber with pressure *Pr*, port 2 is connected to the supply pressure *P* high, and port 3 is connected to the tank. The orifice between the clutch chamber (port 1) and the supply pressure (port 2) is determined by the spool position, which is controlled by the input voltage. When there is no control voltage, the spool is kept at the top position by the return spring. At this position, port 3 is connected to port 1. Therefore, the fluid inside the clutch chamber will flow to the tank.

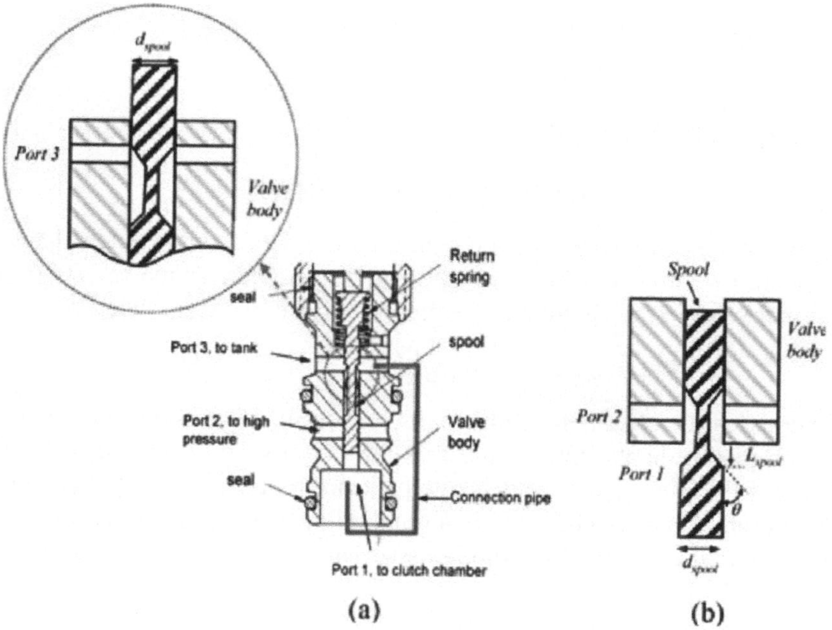

Fig. 2: (a) Spool at the Tank Port,
(b) Spool at the Clutch Chamber Port

When a positive voltage is exerted on the magnetic coil, the induced magnetic force will push the spool toward port 1. As the spool connects port 1 to port 2, the high-pressure fluid will flow into the clutch chamber. The increased pressure in port 1 will push the spool upward and eventually close the orifice between port 1 and port 2. Clearly, the spool position is determined by the magnetic force F_{mag}, the returning spring force F_{spring}, and the chamber pressure *Pr*.

The Clutch Actuation dynamics can be described as follows:

L_{spool} is the spool position, v_{spool} is the spool velocity, M_{spool} is the spool mass, K_{spring} is the spool spring constant, A_{spool} is the cross-sectional area of the spool, and D_{spool} is the damping coefficient. $L_{preload}$ is the spool position, where the orifice between port 1 and port 2 is just closed. F_{mag} is the magnetic force, which is determined by the coil magnetic constant *Kf* and the current *i*. The current *i* is generated by the power amplifier and can be calculated using the input voltage Vol. i_{max} and i_{min} are the maximum and minimum current that can be generated from the power amplifier, and Volmax is the maximum control voltage corresponding to i_{max}.

$$\dot{L}_{spool} = V_{spool} \quad \ldots (1)$$

$$\dot{v}_{spool} = \frac{1}{M_{spool}} \{F_{mag}(Vol) - K_{spring}(L_{spool} - L_{preload}) - D_{spool}v_{spool} - A_{spool}P_r\} \quad \ldots (2)$$

$$F_{mag}(Vol) = K_f \times i = K_f \times \left(\frac{i_{max} - i_{min}}{Vol_{max}} Vol + i_{min}\right) \quad \ldots (3)$$

When the orifice between the high-pressure port and the clutch chamber is open, the fluid will flow into the clutch chamber, which results in the pressure rise in the chamber. The pressure dynamics is modeled as,

$$\dot{P}_r = \frac{\beta(P_r)}{V} [Q(L_{spool}, P_r) - A_p x_2] \quad \ldots (4)$$

where *V* is the chamber volume, which is assumed constant due to small clutch piston displacement, and β is the effective bulk modulus. As the clutch fill typically occurs in a low pressure range (1.68–1.97 bar) while the clutch engagement occurs in a much higher range, the air entrained in the oil could cause the bulk modulus variation.

$$Q(L_{\text{spool}}, P_r) = \begin{cases} \text{sign}(P_h - P_r)C_d\sqrt{\dfrac{2|P_h - P_r|}{\rho}}A_{\text{orifice}}(L_{\text{spool}}), & L_{\text{spool}} > 0 \\ 0, & L_{\text{spool}} = 0 \\ -C_d\sqrt{\dfrac{2|P_r|}{\rho}}A_{\text{dump}}(L_{\text{spool}}), & L_{\text{spool}} < 0 \end{cases} \quad \ldots (5)$$

The dynamics of piston motion can be modeled as:

$$\dot{x}_1 = x_2 \quad \ldots (6)$$

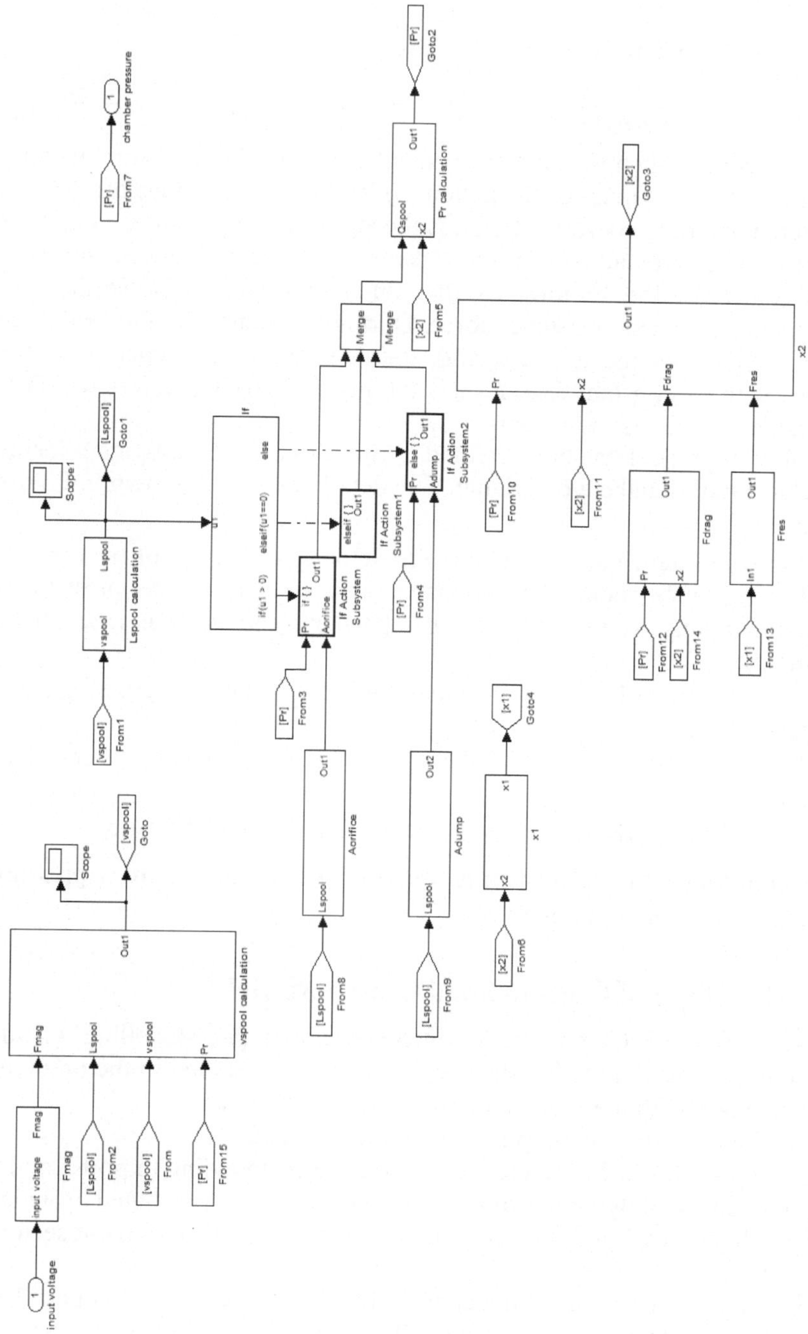

Fig. 3: Simulink Model

$$\dot{x}_2 = \frac{1}{M_p} \times \{A_p \times (P_r + P_c - P_{atm}) - D_p x_2 - F_{drag}(P_r + P_c, x_2) - F_{res}(x_1 + x_{p0})\} \quad \ldots (7)$$

where x_1 is the clutch piston displacement, x_2 is the clutch piston velocity, *Mp* is the effective mass of the piston, *Ap* is the piston surface area, *Dp* is the clutch damping coefficient, P_{atm} is the atmospheric pressure, *xp* 0 is the return spring preload. *Pc* is the centrifugal force induced pressure generated from the rotation of the clutch assembly. F_{res} is the displacement-dependent resistance force. During the clutch fill, the resistance force comes from the return spring; thus, the force F_{res} depends on the spring stiffness constant *K*cs. F_{drag} is the piston seal drag force, which is dependent on the piston motion.

Following model has been created in Simulink. Each subsystem inside calculates individual state variable. The detailed block is shown in Figure 3 where:

- The subsystem v_{spool} calculates the velocity of the spool movement. The v_{spool} subsystem takes the magnetic force and chamber pressure as main inputs. Along with chamber pressure and v_{spool} as a feedback input.
- The subsystem L_{spool} has an integrator block which integrates v_{spool} to give L_{spool}.
- The L_{spool} value is used in the If-Else block to eventually calculate Q_{spool}.
- Finally the Q_{spool} value is used to calculate chamber pressure.

Overall this entire subsystem is used to calculate the chamber pressure from the given input voltage.

2.2 Modelling of Consolidated Clutch Model

The consolidated Clutch model includes Simscape Modelling of the hydraulic actuator, imitating the spool dynamics explained in the previous section, along with the modelling of a friction wet clutch.

The Figure 4 shows the Simscape model created.

The figure shows 2 variable area orifices, corresponding to port 2 and 3 explained in spool dynamics of previous section. The input signal provided to the Subsystems 1 and 2 is an imitation of the L_{spool} displacement seen in Figure 6.

The supply pressure is maintained at 15 bar to port 2 whereas port 3 is connected to atmospheric pressure. The flow rate from port 2 and 3 governs the chamber pressure of the Hydraulic Chamber, which in turn sends the pressure signal to the clutch. The Clutch is provided with a torque source of 30 Nm and the dynamics are observed.

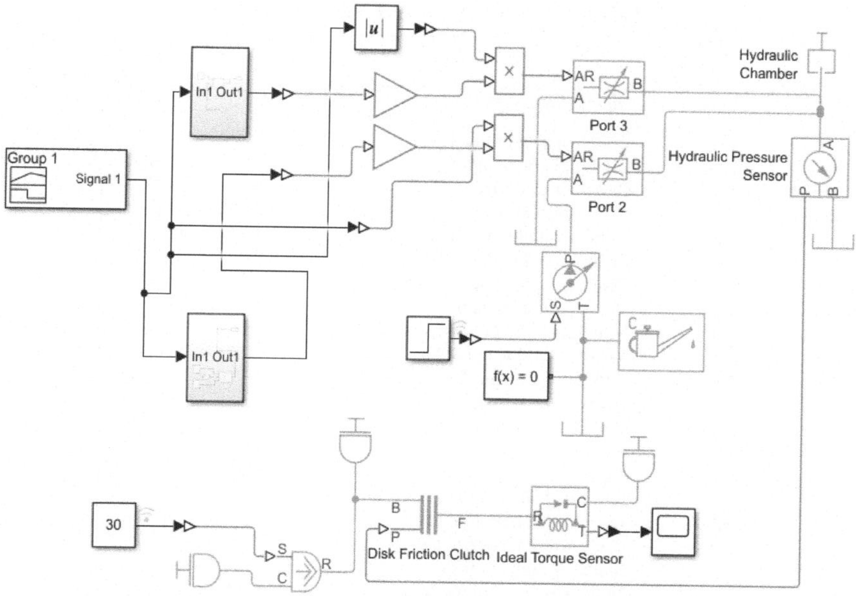

Fig. 4: Simscape Model for Clutch Dynamics

3. RESULT AND ANALYSIS

3.1 Clutch Actuation Model

The following values were used as constants:

Input: The input as shown in Figure 5 where we are varying the voltage in the range of 0 V to 10 V. We start increasing the voltage at 5 second to start the clutch fill process.

Output for Simulation: L_{spool} is the spool displacement. For the given input, we see our L_{spool} starts moving at time 5 second as shown in Figure 6. The L_{spool} value depend directly on the value of the velocity of the spool which in turn depends on the input voltage.

V_{spool} indicates the velocity of the spool. It depends on the value of Chamber Pressure, magnetic force (which depends on input voltage), the feedback V_{spool} and on L_{spool}. The value as shown in Figure 7 suddenly jumps at 5 seconds, takes a dip back and again increases. This value of V_{spool} shows us that the spool is moving in both forward and backward direction with a variation in the velocity, in sync with our input. The

V_{spool} governs the value of L_{spool} which in turn helps us determine the value of the flow rate of the fluid in our system.

Table 1: Parameters for Clutch Actuation

Symbol	Name	Value
M_p	clutch piston mass	0.4(kg)
x_{p0}	Clutch spring preload	1.5928 (mm)
D_p	Clutch piston damping	135.4 (N/m/s)
K_p	Clutch spring stiffness	242640 (N/m)
A_p	Clutch piston area	0.00628 (m^2)
V_0	Clutch chamber volume	4.02e-4 (m^3)
k_m	Piston drag force coefficient	0.001517 (m^2)
c_m	Piston drag force coefficient	5.22 (N)
K_{spring}	Valve spring stiffness	1087.6(N/m)
A_{spool}	Valve spool cross sectional area	5.0645e-006 (m^2)
K_f	Valve magnetic constant	22.7 (Tesla.m)
Vol_{max}	Maximum voltage for valve control	10 (volt)
γ_b	Ratio of specific heats for air	1.4
c_1	Coefficient of air bubble volume variation	0
ρ	Oil density	880 (kg/ m^3)
R	Entrained air content	6%
β_e	Effective bulk modulus of oil	17000 (bar)
k_s	Static friction coefficient	0.00153(m^2)
c_s	Static friction coefficient	5.26(N)
$L_{pre\text{-}load}$	Valve spool spring preload	6.7e-3 (m)
i_{min}	Minimum valve control current	250 (mA)
C_d	Discharge coefficient	0.7
i_{max}	Maximum valve control current	1280 (mA)

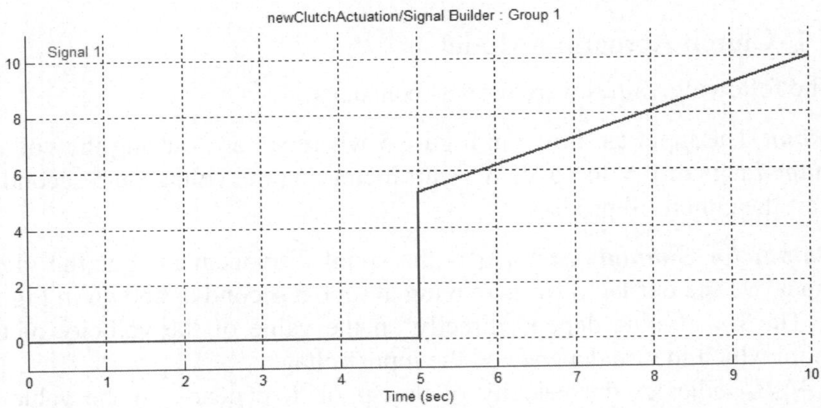

Fig. 5: Input Signal of Voltage Ranging from 0 to 10 V

Q_{spool} is the flow rate of the fluid into the Chamber. It is a function of L_{spool} and Chamber Pressure. Q_{spool} has values mainly in the zone of 0.1 m^3/s at time 5 s, as seen in Figure 8.

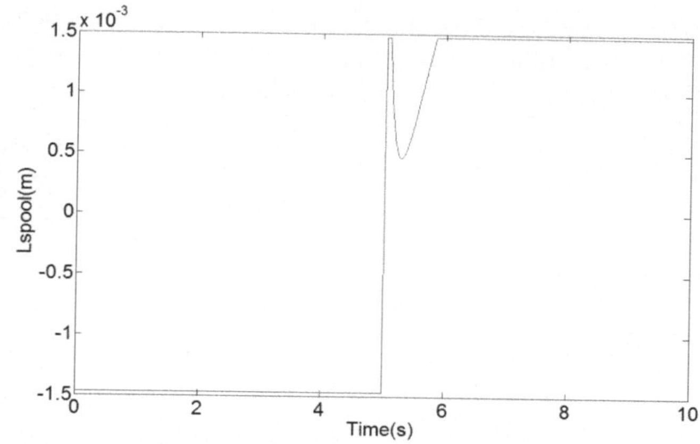

Fig. 6: Output for L_{spool} *vs.* Time

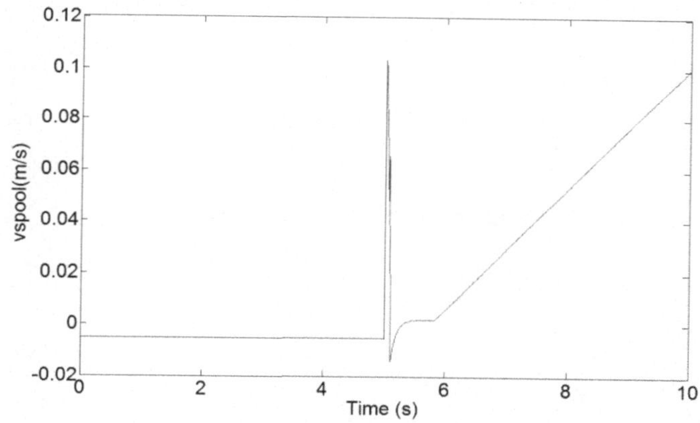

Fig. 7: Output Graph of V_{spool} *vs.* Time

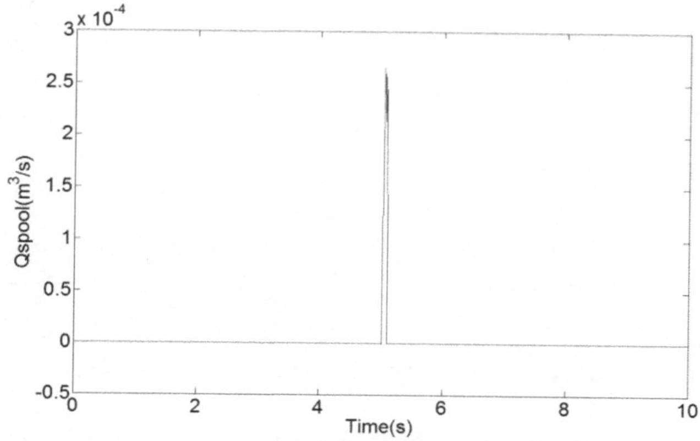

Fig. 8: Output Graph for Q_{spool} *vs.* Time

The graph in Figure 9 shows how the piston position changes with change in the input voltage. The graph shows how the piston position is stable at 0.7 mm after application of voltage at time = 5 s.

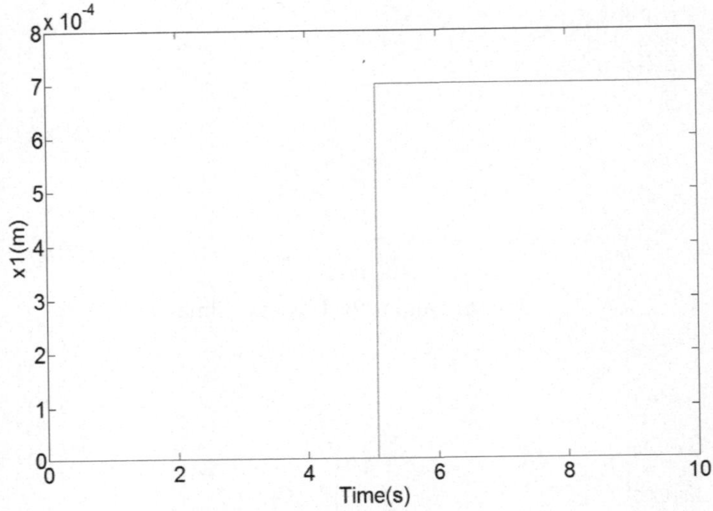

Fig. 9: Graph of x_1 *vs.* Time

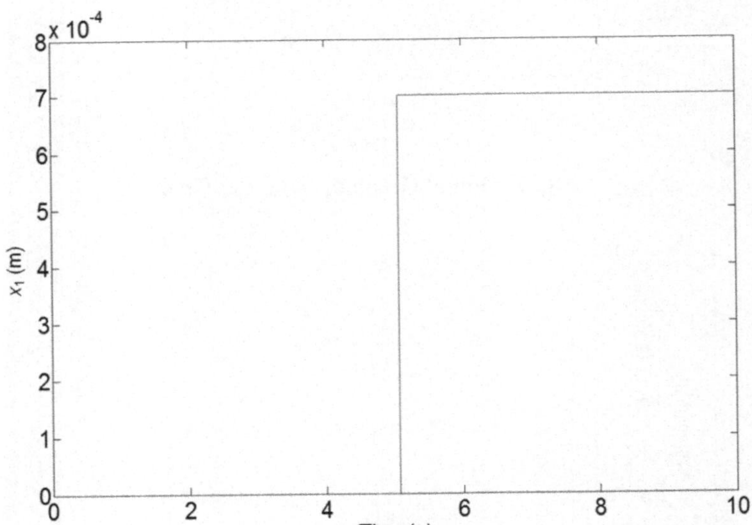

Fig. 10: Graph of Chamber Pressure *vs.* Time

The chamber pressure has a sharp rise after a positive voltage is applied to the system, indicating clutch fill operation successful.

3.2 Dynamic Model of the Clutch in Simscape

The following parameters were used:

Table 2: Parameters for Simscape Model

Name	*Value*
Static friction coefficient	0.12
Kinetic friction coefficient	0.1
Effective radius	0.075 m
Supply Pressure	15 bar
Piston Area	0.00628 m^2

The fluid properties are same as Table 1.

The input signal is an imitation of L_{spool} values shown in Figure 6. The supply torque is 30 Nm. The following results are obtained.

3.2.1 *Chamber Pressure*

This graph shows how the chamber pressure increases to a value of 15 bar from 0 bar as the l_{spool} value varies and opens up the charging orifice while closing the dump orifice. This pressure signal is the one which leads to clutch fill followed by clutch engagement.

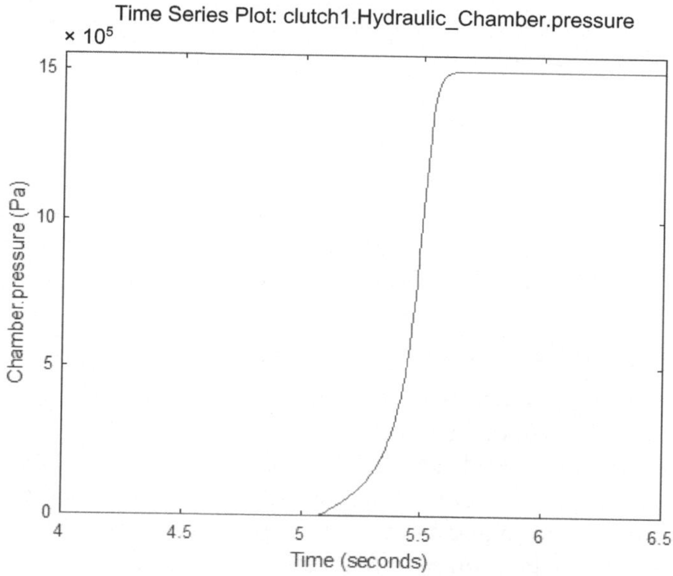

Fig. 11: Variation of Chamber Pressure

3.2.2 *Kinetic Torque and Static Torque*

The maximum torque transferred by the Clutch is the static torque which is given by equation:[6, 9]

$$T_{fmaxs} = \frac{2}{3} R F_n \mu_s N \quad \dots (8)$$

Whereas the kinetic torque, torque transmitted during slipping, is:

$$T_{fmaxk} = \frac{2}{3} R F_n \mu_k N \quad \dots (9)$$

A comparative graph of both these torques are given below in Figure 12. The figure depicts that the static torque of the clutch is 1.2 times of the kinetic torque.

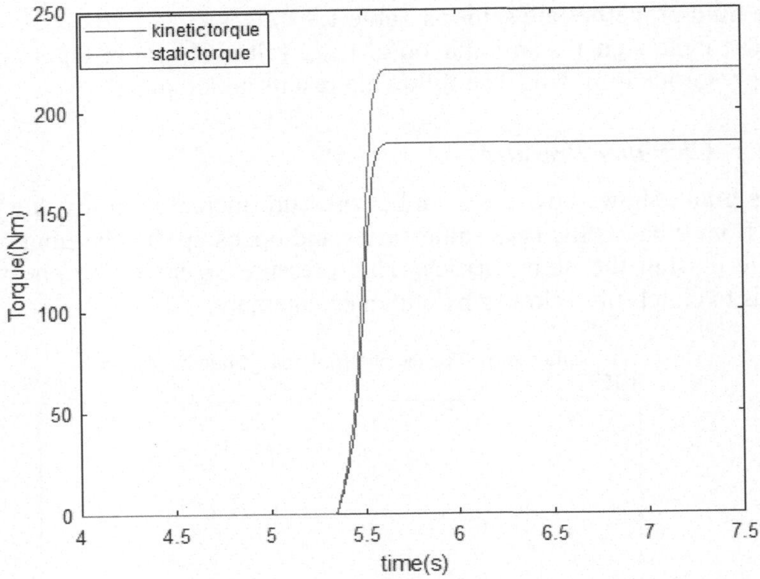

Fig. 12: Comparison of Static and Kinetic Torque

3.2.3 *Angular Velocity on Both Sides of Clutch*

The Clutch engagement process is indicated by the angular velocity of the clutch plates becoming equal.

We see from Figure 11 that the chamber pressure reaches its maximum value at time t = 5.5 s. At the same time we see that the clutch engagement process starts as the engine and vehicle angular speed start moving towards each other. It is only at time t = 6.4 s that complete engagement of clutch happens and the clutch is locked.

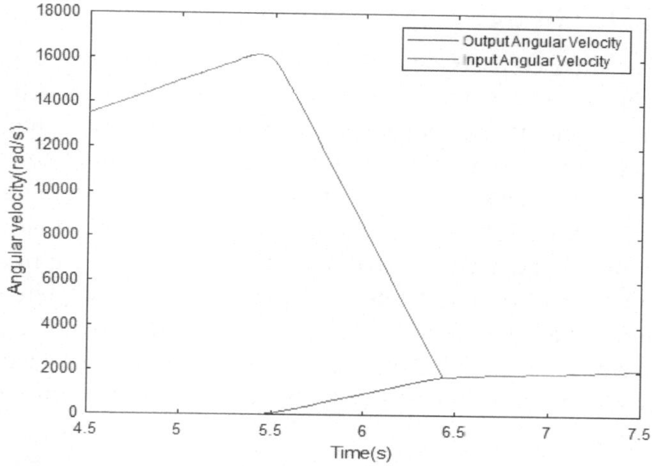

Fig. 13: Angular Velocity on Both Side of Clutch Plate

3.2.4 *Transferred Torque*

The transferred torque is the value of torque which is transferred through the clutch during engagement and disengagement process. The transferred torque has been shown in Figure 14. We observe a spike in the torque at the time of clutch engagement at time t = 5 s, after which it settles to an offset value of 26 Nm. This is indicative of the spike in torque required when we are starting the vehicle from zero speed. Further it settles at an offset value of 26 Nm as some input torque is lost in mechanical losses.

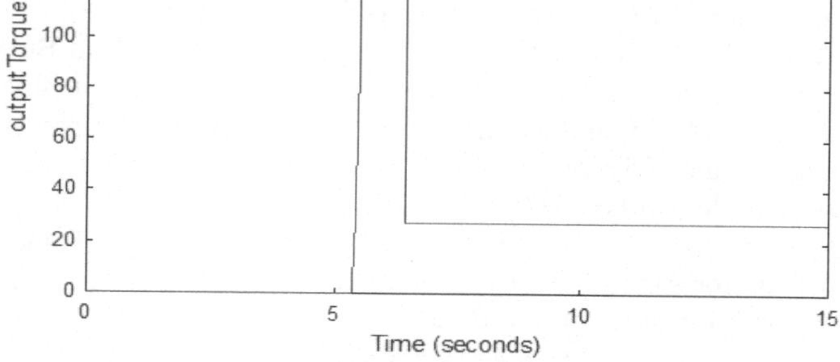

Fig. 14: Transferred Torque

4. CONCLUSION

The work mainly consists of extensive study and implementation of the present work done on clutch dynamics modelling and clutch actuation system modelling. It also includes creating the Simulink model of the Clutch Actuation System. The clutch actuation model is developed by understanding the dynamics of the mechanical and hydraulic subsystem and the spool dynamics. A Simulink model of the clutch actuation system is built to imitate the real time working of the systems. To further analyse our system, Simscape has been used to model the complete actuation system along with the clutch block. This tool enables us to run different test cases and analyse the dynamics of the system in an accurate way, thereby providing an efficient alternative to actually performing tests on the hardware.

REFERENCES

[1] White, E.H., "An Illustrative Look at Energy Flow Through Hybrid Powertrains for Design and Analysis," Blacksburg, Va, 2014.

[2] Sanchawat, A., Agarwal, R., C.R. and B.P., "Simulation of A Drivetrain of a Vehicle Comprising Continuously Variable Transmission," *International Journal of Engineering and Technology,* 5(0975–4024), pp. 3041–3051, 2013.

[3] Chintala, A., Narayana, R.L. and Srinivas, C., "Modelling, Simulation and Control of an Automotive Clutch System," *International Journal of Engineering Research & Technology (IJERT),* 2(12), pp. 3422–3427, 2013.

[4] Sun, X. Song and Zongxuan, "Pressure-Based Clutch Control for Automotive Transmissions using a Sliding-Mode Controller," *ASME Transactions on Mechatronics,* 17(3), pp. 534–547, 2012.

[5] Crosby, H.E. Sarah and Elkobrosy, Galal, "Mathematical Modeling of Continuously Variable Transmission (Cvt) System," *International Journal of Engineering Research & Technology (IJERT),* 5(2), pp. 53–64, 2016.

[6] Quang and Nguyen, H., "Dynamic Modelling of Friction Clutches and Application of this Model in Simulation of Drive Systems," *Periodica Polytechnica Ser. Mech. Eng.,* 42(2), pp. 169–181, 1998.

[7] Pollak, R. Fischer, Küçükay, F., Jürgens, G., Najork, R. and Burkhar, The Automotive Transmission Book, Springer International Publishing, 2015.

[8] Deshbhratar, M.V.J. and Kakde, M.N.U., "Design and Structural Analysis of Single Plate Friction Clutch," *International Journal of Engineering Research & Technology (IJERT),* 2(10), pp. 3726–3732, 2013.

[9] Vasca, F., Iannelli, L., Senatore, A. and Reale, G., "Torque Transmissibility Assessment for Automotive Dry Clutch Engagement," *IEEE/ASME Transactions on Mechatronics,* pp. 1–10, 2011.

[10] Pisatur and Mario, *Dry Clutch for Automated Manual Transmissions. Structural Analysis and Control Strategies,* 2014.

[11] Xie, F., Wang, J. and Wanga, Y., "Study on Cvt Ratio Tracking Controller," In *Elsevier Ltd.*, Changchun, 2010.

[12] Pesce, Thomas, Lienkamp and Markus, "Definition and Optimization of the Drive Train Topology for Electric Vehicles," *World Electric Vehicle Journal,* 5 (2032–6653), pp. 24–36, 2012.

[13] Dalsania, B., Patel, K., Gabani, V., Singhal, A.S. and Dani, V., "Designing and Manufacturing of Continuously Variable Transmission (CVT)," *International Journal of Advances in Production and Mechanical Engineering (IJAPME),* 2(2394–6202), pp. 20–34, 2016.

[14] Băţăuş, Marius, Maciac, Andrei, M. Oprean and Vasiliu, N., "Automotive Clutch Models for Real Time Simulation," *Proceedings of the Romanian Academy,* 2011.

[15] Farkas, Z. and Kerényi, G., "Power Flows and Efficiency Analysis of Out- and Input Coupled Ivt," *Periodica Polytechnica Mechanical Engineering,* 53(2), pp. 61–68, 2009.

[16] Ehsani, M., Gao, Y., S.E. and Gay, A.E., Modern Electric, Hybrid Electric, And Fuel Cell Vehicles: Fundamentals,Theory, And Design, Crc Press, 2004.

[17] Raikwar, R.A., Rewatkar, S. and Kumar, S., "Study the Design Aspects and Function of Wet Type Multi-Plate Clutch," *International Journal of Science Technology & Engineering*, Vol. 2, No. 4, pp. 86–90, 2015.

A Case Study to Development and Validation of Battery Management System through ECU Virtualization

Mistry Pratik Girishchandra[1], **N. Sivakumaran**[2] **and Ritesh Goyal**[3]

[1]Department of Chemical Engineering, National Institute of Technology, Tiruchirapalli, India

[2]Department of Instrumentation and Control Engineering, National Institute of Technology, Tiruchirapalli, India

[3]PES, KPIT Technologies Ltd., Bengaluru, India

E-mail: [1]pratikmistry1994@gmail.com; [2]nsk@nitt.edu; [3]Ritesh.Goyal@kpit.com

ABSTRACT: *Virtualization of the Electronic Control Unit (ECU) is becoming predominant within the automotive industry. It adds to test bench verification of ECU software, which in turn reduces development costs and deployment time. This paper discusses about benefits of virtualization and use cases. Various virtualization techniques and tools can be adopted in order to aid the development and testing of ECU software. The study is on various methods of hardware emulation, where focus is on the design of a solution for development and testing software on a virtual environment. An optimal technique is required for the software testing developed on a particular software architecture, which is scalable across platforms and has standardized architectures. Virtualization of battery management system is done on one of the virtualization tool chain.*

Keywords: Electronic Control Unit, Virtualization, Virtual Validation.

1. INTRODUCTION

With the increasing complexity of software, next generation car manufacturers and their suppliers are confronted with the need to develop models of the electrical and physical parts of the car at different levels of abstraction. There are several phases of abstraction.

- MiL where the functional behavior is often described on a functional level, which is commonly designed in MathWorks Simulink.[8]
- SiL where blocks of the functional design are replaced by actual code that represents the behavior.[8]

…n the Loop' (PiL) where code is compiled for the actual …are. The target hardware communicates directly with the …n 'non-real time'.[8]

- … to PiL, but the target hardware communicates to a PC, …cally equipped with I/O boards to connect to the … of the ECU. The plant model is compiled to execute … computer.[8]

…ers a virtual Hardware in the Loop (vHiL) solution, …ation of the ECU and coupling it with external …, those simulators execute the physical/mechanical …odel (e.g. Mathworks/Simulink). On top of that, the …bly tool Virtualization Studio supports seamless transi-… MiL/SiL to vHiL. Parts of the Simulink model, which are …ated in subsystems, can easily be migrated to the ECU while … of the plant model (e.g. chassis or gearbox) remain in the functional …omain.

2. VIRTUALIZATION AND VIRTUAL ECU

2.1 The AUTOSAR Software Architecure

The AUTOSAR standard defines an architecture as shown in Figure 1 for embedded software on ECUs. The AUTOSAR software architecture is a layered architecture that has the goal to abstract from hardware components. AUTOSAR systems can be divided in six main components.

- Application Software is the software implementing the unique features of an ECU, e.g., the behavior of the Cell Balancing.[5]

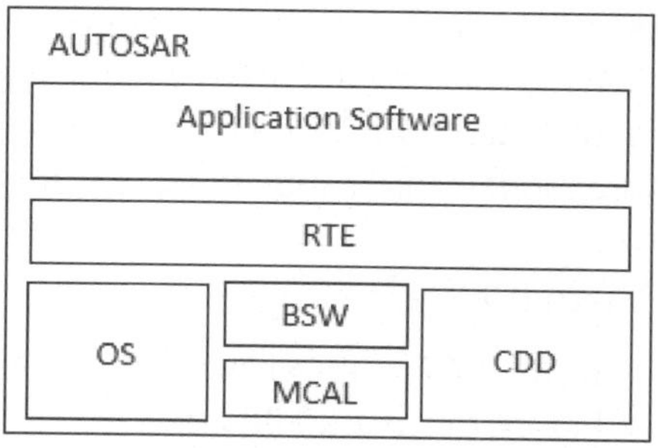

Fig. 1: AUTOSAR Architecture

Cell Balancing: This module monitors the fast charger states, cell voltages, die temperatures, and decides on the cell balancing switch states during fast charging.

- Runtime Environment is the communication layer, whic the signals directly between ASW components or the software's (BSW) communication stack. The RTE wi dispatch the data from one component directly to another if they are deployed on the same ECU, or the data is se communication stack in the base software.[5]
- Base Software is software that provides basic functionality of a Typical software components are communication stacks su CAN, automotive Ethernet, or flexray.[5]
- Operating System is a real-time system that is responsible for exe ing code at the right time and with a defined maximum duration.[5]
- Microcontroller Abstraction Layer is the driver layer and specific fo the used ECU hardware.
- Complex Device Drivers contain special code which is not commonly used and this not part of the AUTOASR specification.

2.2 Category of Virtual ECU

vECUs can be roughly classified into three categories:

- vECUs that contain only the ASW and RTE (and optionally an OS). This is useful for quick testing of basic functions of the ASW without using base software components like communication.[5]
- vECUs that consists of ASW, RTE, BSW, OS and a virtual MCAL (for x86 systems). A more realistic behavior of the real ECU can be simulated with this kind of vECU. The scheduling of tasks is considered and the functionality of the BSW can be tested.[5]
- If a more realistic behavior of the vECU is desired, virtual hardware can be used. All software components of the real ECU can be re-used, including the target MCAL layer. The MCAL is used with detailed hardware models that simulate real timing behavior.[5]

2.3 Benefits of Virtual ECU

- *Early Integration of ECU:* In virtual ECU, we are able to do integration even before the availability of hardware.
- *Reduction of Hardware Costs and Increased Hardware Efficiency:* Without connecting hardware to our system, we can measure the data. There is no connection because of that we can reduce the cost of connection and by the same time we will get accurate output from system itself.

- *Complexity:* With the use of virtualization, any kind of complex hardware we can emulate on software and without connecting any hardware, we can virtually validate system.
- *Virtualization Offers Opportunities to Test New Ideas:* Since we have not connected any hardware to our system, it will not harm the software or hardware. Therefore, we can run and test many ideas on virtualization.

3. VIRTUALIZATION OF ECU IN DSPACE TOOL CHAIN

3.1 Brief Idea about the Virtualization Tool Chain

Basic procedure to carry out for building of virtual ECU in any of the software tool chain.

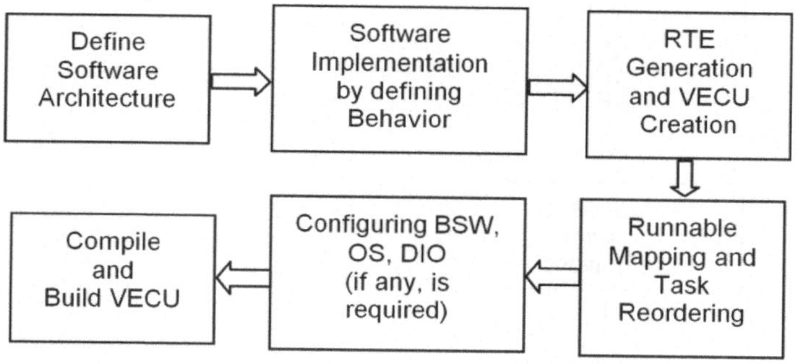

Fig. 2: Common Work Flow for Creation of VECU

There are many tool chains is available for virtualization of an ECU such as ETAS (ISOLAR), Qtronic (Silver), DSpace (System Desk). Each tool chain has its own merits and shortcomings.

- *DSpace (System Desk):* In this software we can do the virtualization by the use of *.c and *.h behaviors files. Those files are consist about the functionality of application software component.
- *Qtronic (Silver):* This software has some more advantage over DSpace. Silver has a benefit to do virtualization with the use of HEX file. If we are using external chip for system then emulation of is also possible in Silver software.

In upcoming section, workflow for DSpace has been discussed. Virtualization of battery management is done on DSpace tool chain and virtually validated component's signal by running open loop simulation.

3.2 Virtualization in DSpace Tool Chain

There are three steps for virtualization and validation of an ECU in DSpace.

1. *System Desk:* A system desk project holds all the elements, configurations and information that are required to carry out a specific task.
2. *VEOS Player:* System Desk allows you to build V-ECUs to simulate them with VEOS and provides support for debugging target code.
3. *Control Desk:* A control desk project consist of all the data with harness model, where simulation and measurement of parameters is possible.

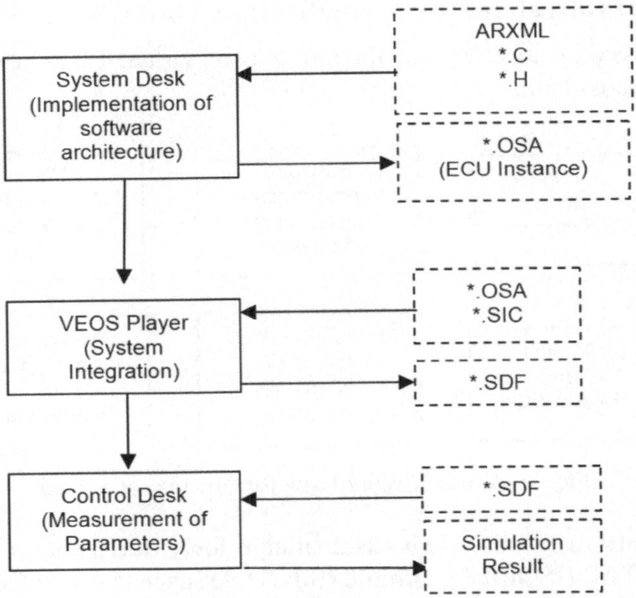

Fig. 3: Work flow for DSpace Tool Chain

3.3 Virtualization of BMS in Windows Enviornment

Battery Management System Overview: In Electric Vehicles, traction power is generated by electric motor drive system. The electric power needed to drive the traction motor is supplied by high voltage and high-power secondary battery pack. This high voltage battery pack typically consists of several cells connected in series and parallel combination. It is important that battery cells are loaded and charged in the specified operating region so that performance, battery life and safety goals are met. This is achieved by Battery management system (BMS). BMS is an

ECU that consists of one or more PCB with analog, digital circuitry, micro controller, sensors, contactors, electrical harness, communication ports and protective enclosure.

The following are functions of BMS:

1. *Battery monitoring:* Cell voltages, cell current, cell temperatures.
2. *Battery state estimation:* SOC, SOH, Peak Power capability.
3. *Battery protection:* Operating contactors in case of faults, battery overloading or misuse.
4. *Cell equalization:* Balances cells connected in series so that battery works up to its maximum capacity.
5. *Thermal management:* Controls temperature of battery with the help of water pump or heater for optimum battery use.

SOC: There are three options to calculate the State of Charge of battery pack (SOC) which are listed below:

- *Coulomb counting method (Direct):* This method provides very accurate measure of current flow but this method cannot precisely account for losses of energy internal to battery.
- *OCV vs. SOC method (Indirect):* This method can do accurate estimation of the SOC depending upon the accurate measurement of OCV, but cannot be used for dynamic SOC Estimation.
- *Combination of Direct and Indirect Method:* By employing both Direct and Indirect methods in appropriate conditions one at a time, the drawbacks of both methods can be overcome.

By considering the above points, we have finalized to use the Combination of Direct and Indirect Method because by using this method efficient and high accuracy of estimation can be obtained. Combination of Direct and Indirect Method automatically gives a dynamic estimate of the SOC and its uncertainty. It automatically provides dynamic error-bounds on the estimates. In this method, 70% of sub-systems developed for SOC can be reused in SOH estimation. This method is little complex due to a set of recursive equations that are repeatedly evaluated as the system operates.

Method 1: Coulomb Counting

$$SOC(n) = SOC(n-1) + \frac{i \times \Delta T}{C}$$

Method 2: OCV Based

$$SOC(n) = f(OCV(t))$$

Method 3: IIR Filter

$$SOC_{Final(n)} = (1 - K) \times SOC_{Final(n-1)} + K \times SOC(n)$$

SOH: Pack level SOH value for the BMS. Capacity fade estimation: Internal resistance of battery pack is used to calculate open circuit voltage (OCV) of the battery. Calculated OCV is used to calculate SOC. Difference in SOC and coulomb count transferred is used to calculated capacity fade. Internal resistance of battery pack is calculated as sum of internal resistance of cells in battery pack. OCV also being assumed as terminal voltage when current value is very low for particular time.

3.4 Workflow for Virtualization of BMS

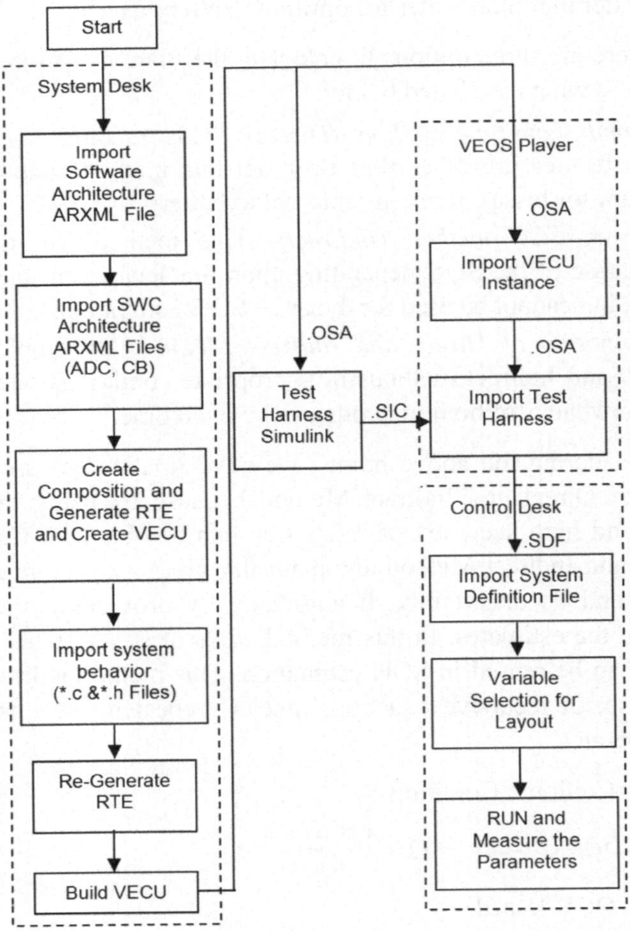

Fig. 4: Work Flow for Virtualization of BMS

3.5 Work Flow for Unit Testing ADC Component in BMS

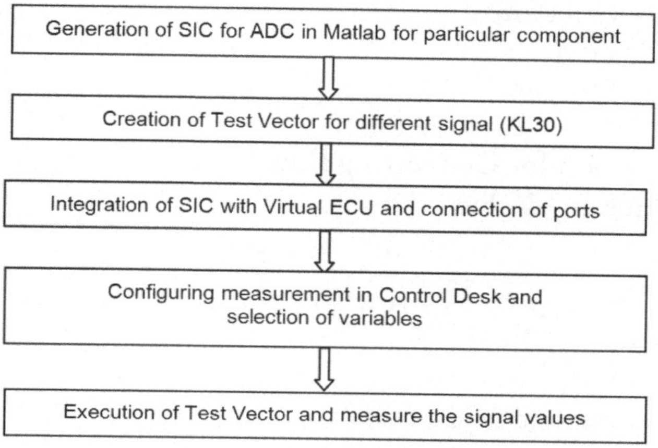

Fig. 5: Unit Testing Work Flow

3.6 Signal Flow in Virtual ECU

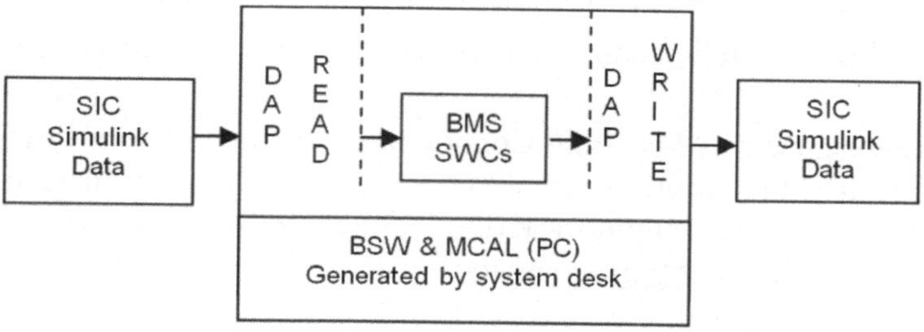

Fig. 6: Signal Flow from SIC to SWCs

Here as shown in Figure 6, Stimulus is given to simulink. From SIC signal is flowing to data access point read and from there it will go to BMS Software components where logic is created according to system requirements. Final output is coming from DAP write and it is again going back to simulink model.

4. RESULTS AND CONCLUSION

Unit Testing of ASW_ADC: ASW_ADC has following RPorts and PPorts, which is getting input from INF_ADC and giving output to other component such as Cell Balancing, Battery Current Sensing, Cell Voltage Acquisition, etc.

1. KL15 Voltage
2. Negative Coil Current

3. Pre-charge Coil Current
4. Positive Coil Current
5. KL30 Voltage
6. Ignition Wake up.

4.1 Calculation for Converting Raw Value to Engineering Value

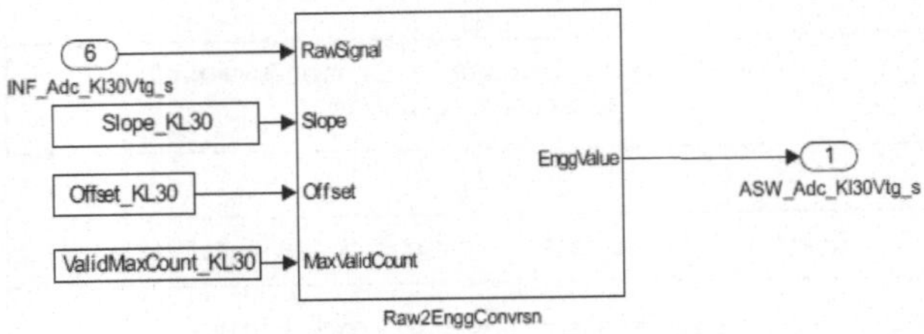

Fig. 7: Simulink Diagram of ASW_ADC for KL30 Voltage

(Slope × Raw Signal) + Offset = Engg value

Here, Figure 7 tells about the conversion of raw signal to engineering value.

If Raw signal <= Max Valid Count condition satisfies then it will give engineering value according to above equation or else it will give delay with previous value. Output for KL30 is shown in Figure 8.

Constants: Slope_KL30 = 0.0049

Offset_KL30 = 0.0

ValidMaxCount_KL30 = 3989

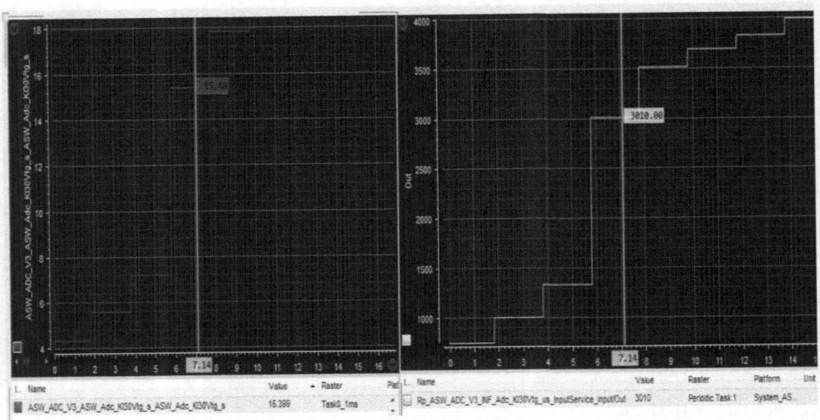

Fig. 8: Simulation Result for KL30 Voltage

Table 1: Input and Outputs for KL30 Voltage

Time	*0–2*	*2–4*	*6–8*	*8–10*	*10–12*	*12–14*
Rp	751	1005	3010	3510	3690	3826
Pp	4.3	5.57	15.4	17.8	18.5	18.5

4.2 Result for Chip Die Temperature

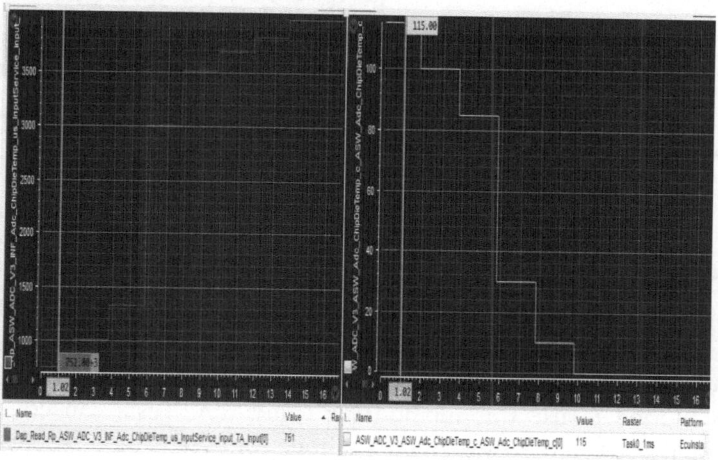

Fig. 9: Simulation Result for Chip Die Temperature

Table 2: Input and Outputs for Chip Die Temperature

Time(s)	*0–2*	*4–6*	*6–8*	*10–12*	*12–14*
Rp	751	1335	3010	3690	3826
Pp (Deg C)	115	85	30	0	0 (error due to dt)
Expected Temp	115	85	30	0	–15

During the testing of chip die temperature came to know that it is not giving the value as expected as shown in figure 9 and Table 2. So in the process of finding the reason found that it is due to data type error in Application software's Simulink model. So like that, we can test the signal virtually without connecting it to hardware.

4.3 Result for Cell Balancing in Different State

According to functionality created in simulink we are getting output as shown in Figure 10 for Cell balancing state as 1. Where 1 represents Cell Balancing is in progress. So like that we can do the early validation of system in virtual enviorment.

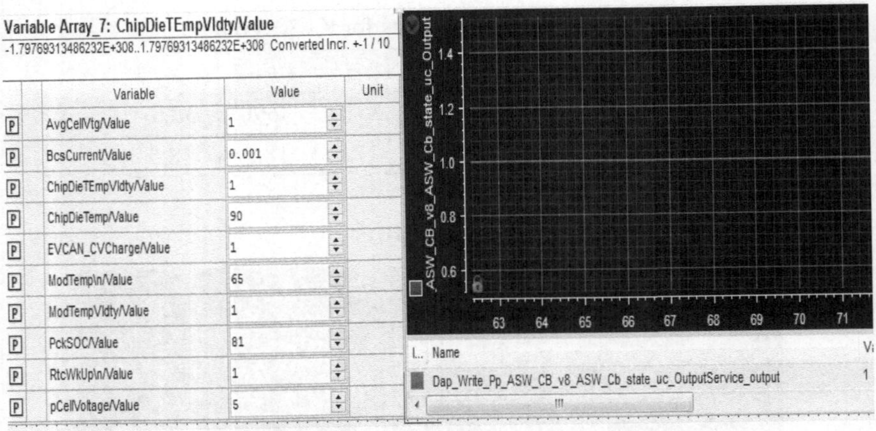

Fig. 10: Cell Balancing in Control Desk

5. CONCLUSION

Virtualization technology is suitable to verify and virtually validate car's ECUs and their communication. Virtualization plays a major role in verification and early validation of ECUs that is why many industries has aleady started to use this method for validation. Tests can cover, e.g., functional behavior, software integration, communication, security, etc. Simulink based Code of selected tasks of an ECU executed by the virtual ECU tool DSpace on Windows PC, open loop testing done by measurements in control desk. This kind of simulation opens new possibilities to move development tasks to PCs, where they can be processed faster, cheaper or better, without requiring access to the underlying C code. With the help of plant model closed testing is also possible in virtual ECU.

Abbreviations and Acronyms

- ARXML : Autosar Extensible Markup Language
- AUTOSAR : Automotive Open System Architecture
- A2L : ASAM MCD-2 MC Language
- C & h : C Code file and header File
- OSA : Offline simulation application
- RTE : Run time Environment
- BSW : Basic software
- FMU : Functional Mock-Up Unit
- SIC : System Information Container
- SDF : System Definition file
- BMS : Battery Management System

- Rp : Receiver Port
- Pp : Provider Port
- SOC : State of Charge
- SOH : State of health
- DIO : Digital Input Output
- ASW : Application Software
- OS : Operating System
- MCAL : Microcontroller Abstraction Layer

REFERENCES

[1] ISOLAR-EVE_V3.2.1_UsersGuide.

[2] Reinhardt, D., Kaule, D. and Kucera, "Achieving a Scalable E/E-Architecture Using AUTOSAR and Virtualization," *SAE Int. J. Passeng. Cars – Electron. Electr,* No. 10.4271/2013-01-1399, Syst. 6(2):2013.

[3] Katyal, R.K. and S.S., "Virtualization for ECU Platform Software Testing in Automotive Embedded," *general Motors Bangalore.,* December, 2011.

[4] Coelingh, E., Chaumette, P. and Andersson, M., "Open-Interface Definitions for Automotive Systems—Application to a Brake-by-Wire System," *Technical Paper 2002-01-0267, SAE In-ternational,* March 2002.

[5] Pelzl, J., Wolf, M. and Wollinger, Thomas, escrypt GmbH, "Virtualization Technologies for Cars," *Embedded Security.*

[6] Rink, A., Chrisofakis, E. and Tatar, M., "Automating Test of Control Software—Method for Automatic TestGeneration," *ATZelektronik,* Vols. 4, pp. 24–27., 6/2009.

[7] DSpace, "System Desk Manual from DSpace Software's Help tab".

[8] N.S., N.P. and S.S. and Garrelfs, R., "MIL/SIL/PIL Approach A new paradigm in Model Based Development," in *Continental AG.*

Mobile SCADA Solution for Condition Monitoring of Conveyor Systems

T. Shuprajhaa[1], K. Srinivasan[1*], G. Saravana Ilango[1] and M.P. Ambalavanan[2]

[1]National Institute of Technology, Trichy

[2]DGM/CARD/NLCIL

*E-mail: Srinikkn@gmail.com

ABSTRACT: *Internet of Things (IfoT) is one domain which has its deep-rooted applications in almost all areas of engineering. Implementation of advanced smart monitoring in industries is the need of the hour. A conveyor belt system is normally consisting of many critical process parameters that need to be monitored and controlled at specified threshold levels. Condition monitoring of the conveyor system enables the user to have full detailed inspection of the overall system. NLCIL mines have movable conveyors which demand constant monitoring and control. SCADA enabled monitoring is developed for the prototype setup available in NITT and in-depth condition monitoring is enabled for the same. A mobile SCADA solution for the condition monitoring of conveyor system has also been proposed.*

Keywords: IoT, Sensors, Data Acquisition, Mobile SCADA.

1. INTRODUCTION

With the advent of internet, the domain Internet of things is gaining more and more importance. Enabling monitoring of conveyor drive parameters from a remote location has lot of advantages. A taxonomy aiding in defining the components required for Internet of things has been presented in.[1] When there is a provision to closely monitor the parameters the occurrence of any failure could be detected before and suitable advance prevention techniques could be applied. Machinery malfunctions are usually easily predictable.

Maintenance of the system could also be done online without having any interference with the normal operation of the system. Monitoring the parameters of any system through online mode is highly advantageous as it explicitly ensures the smooth functioning of the system under consideration. The conveyors are large in size and very long in length. The NLCIL conveyors involves many parameters like voltage, current, temperature, vibration and power. All the above parameters are widely scattered in a wide area. Online monitoring of the above parameters is inevitable. A detailed survey of current ongoing research in IoT, key enabling technologies, major applications of IoT in industries is provided in.[2] Application of IoT in bulk solid material handling and transportation system and utilisation of big data has been taken as the case study and has been discussed clearly in.[3] Steps to extract data from big data have been explained in detail. Smart idler concept for continuous monitoring of the belt conveyor system has been presented. RFID sensor nodes are fitted in rolls to monitor the temperature of the rolls. A detailed survey of the existing internet protocols has been presented in.[4] Standardized Protocol Stack for the Internet of (Important) things has been extensively dealt in.[5] Condition monitoring of the environment in residential homes has been proposed in.[6] A Survey on application layer protocols for the Internet of Things is given in.[7] The main concentration of this proposed paper is on developing a mobile SCADA solution for real time condition monitoring of conveyor suitable for NLC environment.

2. MOBILE SCADA BASED CONDITION MONITORING

An overview of NLCIL mines and conveyor network is shown in Figure 1(a) and 1(b). A prototype conveyor setup similar to the ones at NLCIL

(a)

(b)

Fig. 1: (a) Overview of NLC Mines
(b) Overview of Conveyor Network in NLCIL

has been commissioned at National Institute of Technology, Tiruchirappalli. The conveyor system consists of four numbers of induction motor each of 2.2 KW rating. Motor and Gear box vibration has been measured using appropriate vibration sensor. Motor surface temperature and winding temperature is measured using RTDs. The current and voltage of the motors are measured using Energy meter and current and potential transformers. IP camera is installed in the conveyor setup to take photographs at regular interval. The laboratory prototype setup and the sensors mounted on one single motor along with the local controller is shown in Figure 2(a) and 2(b).

(a)

(b)

Fig. 2: (a) Prototype Setup at NITT
(b) Sensors Mounted on Single Motor and Local Controller

NLCIL Mines are operating with conveyor system for the transportation of over burden and Lignite. In a conveyor system, each drive head consists of four slip ring induction motors catering to the full load of the conveyor.

SCADA based monitoring of the conveyor motors measurement parameters through internet could be implemented by following a certain typical network architectural design strategy together with the application of standard communication protocol.

Suitable sensors for measurement of vibration, current, voltage and temperature are installed on the conveyor system. The data from the sensors are acquired through the data acquisition module and sent to the master controller module in the local control station. The overall architecture of the SCADA-based condition monitoring of the conveyor system is as shown in Figure 3. The proposed architecture involves not only condition monitoring of the parameters effectively on-line but also enables effective corrective actions to be taken when needed. Based on the measured signals as input, the controller in the local station has to take appropriate control action. When the measured signals deviate from 6 desired set limit of the parameters, appropriate corrective action will be taken, i.e. it involves bi-directional transfer of signals. Similarly, the design of such bi-directional architecture becomes challenging one due to mobility of the conveyers drive heads, formation of terrains at different levels. Controller will be capable of recording a minimum of 100 process parameters simultaneously. The logged data file shall be compatible in MS excel format or shall have .csv extension.

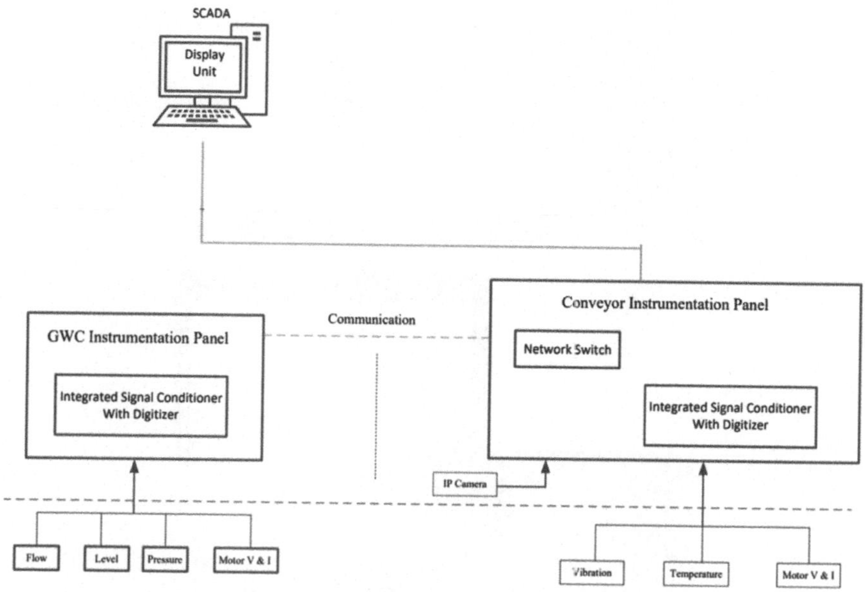

Fig. 3: Architecture of Cloud-Based Monitoring of the Conveyor System

Sensor data of the last 30 minutes will be presented to the user in the screen. A minimum of three-month duration of data will be available in the data logger at the storage for the user to be retrieving it at a later time. Two level of user authentication of data shall be provided. Data logger shall have reliable battery for power backup of minimum 2 hours. Logged data shall be visually presented in numeric, digital and real time graphical format. Provisions for the user to view any specific required sensor data of a specific time duration is also be provided. Excel compatible report generation is feasible. Graphical User Interface screen showing the live status of the conveyor system and its parameters trend is designed. Mobile based application has been created using Java and Python script to view this GUI in mobile phones and tablets. The GUI screen of this mobile SCADA-based monitoring architecture is shown in Figure 4. The graphical representation of the data acquired is shown in Figure 5.

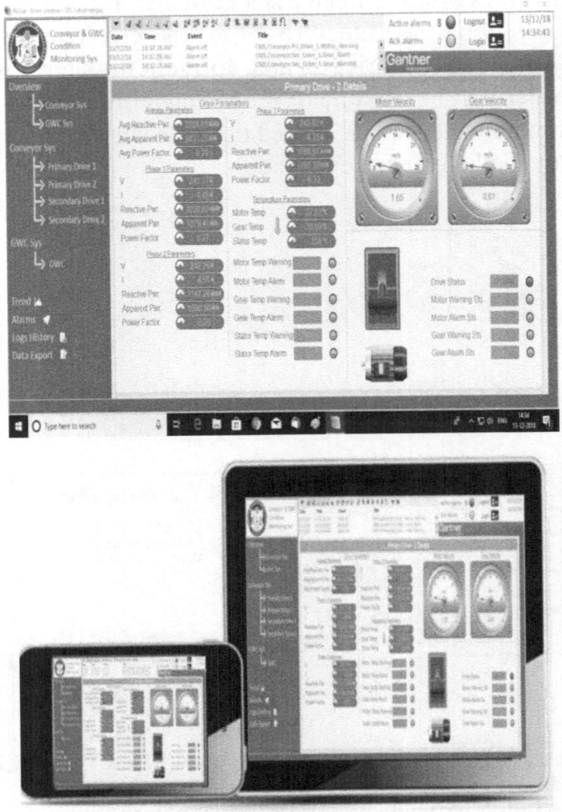

Fig. 4: GUI Screen of Mobile SCADA-based Monitoring Architecture

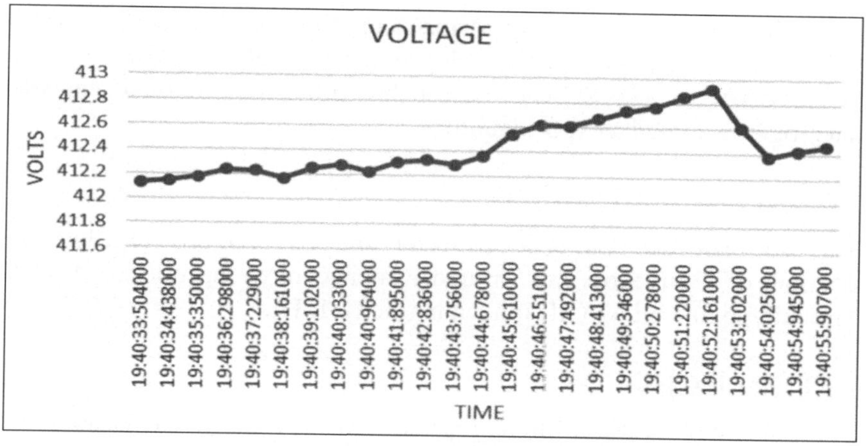

Fig. 5(a): Graphical Representation of Voltage Data Acquired from Energy Meter

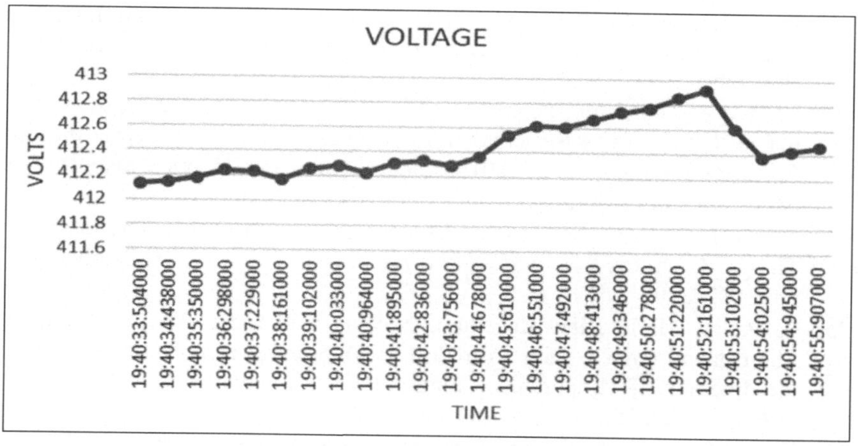

Fig. 5(b): Graphical Representation of Current Data Acquired from Energy Meter

3. CONCLUSION

Intelligent monitoring and control are always required to maintain the reliability and to improve the performance efficiency of the conveyor system. Provisions for local SCADA based monitoring of the conveyor drive motor parameters has been built. Live update of different conveyor parameters has been demonstrated in the prototype setup at NITT. SCADA screen has been developed for monitoring the individual motors in the conveyor. Vibration, temperature, current, voltage, active power, reactive power, total power, power factor data has been acquired from the respective sensors and energy meters and data visualization is enabled. Mobile based app has been developed for ease of visualization of the SCADA screen. The SCADA screen has provisions of trend analysis, data logging and data export facilities.

ACKNOWLEDGEMENT

This work is carried out as a part of the joint research project, Electronification of Conveyor and ground water control in mines of NIT-Tiruchirappalli and NLCIL funded by the CMPDI.

REFERENCES

[1] Gubbi, Jayavardhana; Buyya, Rajkumar; Marusic, Slaven and Palaniswami, Marimuthu (2013). Internet of things (IoT): a vision, architectural elements, and future directions, *Future Generation Computer Systems*, Vol. 29, Issue 7, September 2013, pp. 1645–1660.

[2] Xu, Li Da; He, Wu and Li, Shancang (2014). Internet of Things in Industries: A Survey, *IEEE Transactions on Industrial Informatics*, Vol. 10, No. 4, November 2014.

[3] Lodewijks, G., Li, W., Pang, Y. and Jiang, X. (2016). An Application of the IoT in Belt Conveyor Systems. In: Li W. *et al.* (eds) *Internet and Distributed Computing Systems.* IDCS 2016. Lecture Notes in Computer Science, Vol. 9864. Springer, Cham.

[4] Atzori, Luigi; Iera, Antonio and Morabito, Giacomo (2010). The Internet of Things: A survey, *Computer Networks*, Vol. 54, Issue 15, 28 October 2010, pp. 2787–2805.

[5] Palattella, Maria Rita; Accettura, Nicola; Vilajosana, Xavier; Watteyne, Thomas and Grieco, Luigi Alfredo (2013). Standardized Protocol Stack for the Internet of (Important) Things, *IEEE Communications Survey and Tutorials*, Vol. 15, No. 3, Third Quarter 2013.

[6] Tebje Kelly, Sean Dieter; Suryadevara, Nagender Kumar and Mukhopadhyay, Subhas Chandra (2013). Towards the Implementation of IoT for Environmental Condition Monitoring in Homes, *IEEE Sensors Journal*, Vol. 13, No. 10.

[7] Karagiannis, Vasileios; Chatzimisios, Periklis; Vazquez-Gallego, Francisco and Alonso-zarate, Jesus (2015). A Survey on application layer protocols for the Internet of Things, Transaction on IoT and Cloud Computing, ISSN: 2331-4753 (Print) ISSN: 2331-4761 (Online).

Design of Optimal Multi-Loop Controller for a Class of Linear 2-by-2 MIMO Process

V. Sujatha[1], J.R. Suresh Babu[2], S. Gayathri[3] and Rames C. Panda[4]

[1]Department of Instrumentation & Control Engineering, Sri Manakula Vinayagar Engineering College, Puducherry, India

[2]Wii Tronics Solution Pvt. Ltd. Chennai

[3]Department of Electronics and Instrumentation Engineering, Velammal Engineering College, Chennai, India

[4]Department of Chemical Engineering, Central Leather Research Institute, Chennai, India

E-mail: [1]dr.vijaysuji@gmail.com; [2]sureshmagix@gmail.com; [3]icegayu@yahoo.com; [4]rames.panda@gmail.com

ABSTRACT: *In automatic process control systems, it is important to determine the controller parameters. Optimization of inputs in process control is used to adjust a process to optimize a specified set of parameters at a reduced cost. Hence optimization is done using Levenberg Marquardt algorithm in this work. Application of this algorithm to a real time coupled tanks system is presented in this paper. This research considers an objective function for multi-loop PI controllers which provide minimum area under the closed loop undesired responses. Applications of the proposed method not only provides information on optimization of inputs but also helps to suggest suitable input-output pairing based on least area under the undesired closed-loop responses and accordingly reconfigures the closed-loop system. This application would be very helpful for improving the control performance. The simulation and real time experimental results demonstrate the effectiveness of the proposed monitoring method and redesigning of controllers for linear 2-by-2 MIMO systems.*

Keywords: Closed Loop Undesired Responses, Monitoring, Redesign of Controller, Objective Function, Linear 2-by-2 MIMO Process.

1. INTRODUCTION

The majority of industrial processes are multivariable in nature, that is, there are many output variables to be controlled, and more than one input variable is coupled with the outputs. Multivariable processes are difficult to control due to the presence of the interactions. Such processes are found in process industries as they arise from the design of process units that are subject to rigid product quality specifications, are more energy efficient, have more material integration, and have better environmental performance. Most of the unit operations in process industry require control over product rate and quality by adjusting one/ more inputs to the process; thus making multivariable systems complex. Common practical processes with these features are liquid-level control in coupled tanks, composition/temperature control in top and bottom of distillation columns, concentration/temperature control of CSTR, heat exchangers and flash drums. Mathematical models of undesirable relay response (off diagonal transfer function) can be used as a tool of interaction measure and for subsequent selection of input-output pairs for control. In the presence of accurate interactions, multi-loop controllers designed using BLT method (Yu)[1] can be redesigned by minimizing the area under the closed loop undesired/disturbance responses thus producing multi-objective function.

Distillation is widely used for separating components from mixture in refineries. Composition of top and bottom products are controlled by adjusting energy input to the column. A common scheme is to use reflux flow to control top product composition whilst heat input is used to control bottom product composition. However, changes in reflux also affect bottom product composition and component fractions in the top product stream are also affected by changes in heat input. Hence, loop interactions occur in composition control of distillation column (Deshpande).[2] Control system design needs availability of linear models for the multivariable system (Albertos & Sala).[3]

This paper describes a real time experimental multivariable control systems called coupled tanks system, consisting of four interconnected water tanks and two pumps. The inputs are voltages to the two pumps and the outputs are the water level in the lower two tanks. The objective is to control the level in lower two tanks which are very difficult due to interactions. This needs knowledge in control structure selection or interaction analysis. In the next section, a brief state of art on interaction analysis is presented.

Relative gain array (RGA) (Bristol)[4] is the most discussed method for analyzing interactions and it is based on steady state gain information of MIMO processes. Control loops should have input-output pairs which give positive relative gains that have values which are as close as unity as possible. Because the relative gain array reveals only static interactions in a process, extensions to it have been proposed which include dynamic interactions effects too, for example in Witcher and McAvoy[5] and Tung and Edgar.[6] Niederlinski index (NI) is a useful tool to analyze interactions and stability of the control loop pairings determined using process gain matrix. The values of NI need to be positive. A negative value of NI will imply that the system is unstable. NI is used to check if the system (more than 2×2) is unstable or not. NI will detect instability introduced by closing the other control loops. Generally, NI is not used for systems with time delays. Any loop pairing is unacceptable if it leads to a control system configuration for which the NI is negative. But both RGA and NI do not provide dynamic information on the process transients. They do not give information on change in input/output pairing for instances when there is a sudden load disturbance. Singular value decomposition (SVD) is a useful tool to determine whether a system will be prone to control loop interactions resulting in sensitivity problems that rises from model mismatch in process gains. SVD is dependent upon the scaling of the process model. Lieslehto[7] proposed the scaled gain matrix which is based on the scaling of input and output variables. Though SVD has good geometric interpretation in terms of selection of measurement and pairing of variables, SVD depends on input-output scaling. Moreover, with weak interactions and with large dimensional systems they induce to go for more criteria for selection of pairs. The Morari Resiliency index (MRI)[8] is the minimum singular value *(g)* of the plant transfer function matrix $G(j\omega)$. The set of manipulated variables that gives the largest minimum singular value over the frequency range of interest is the best. The MRI is a measure of the inherent ability of the process (control structure) to handle disturbances, model plant mismatches, changes in operating conditions, etc. The larger the value of MRI, the more resilient is the control structure. Dynamic Relative Gain Array (DRGA) is defined to extend the RGA notion to non-zero frequencies. The RGA provides only limited knowledge about when to use multivariable controllers and gives no indication of how to choose multivariable controller structures. A somewhat different approach for investigating channel interaction was therefore employed by Conley and Salgado[9] and Salgado and Conley[10] when considering observability and controllability gramians in so called Participation Matrices (PM). In a similar approach Wittenmark and Salgado,[11] Ali and Bijan[12] introduced the Hankel Interaction Index Array (HIIA). These gramian based interaction measures seem to overcome most of the disadvantages of the RGA. The use of the system H_2 norm as a base for an interaction measure has been proposed by Birk and Medvedev[13] as an alternative to the HIIA. But, dynamic simulation is a powerful tool to be used to test the viability of a control scheme during various process disturbances.

The main objective of the present work is to capture the disturbance dynamics thereby measure the interaction in terms of the area under the closed loop undesired response. The goal is to minimize the interaction by using medium-scale algorithm with termination tolerances for the step and objective function on the order of 0.001. The optimization[14] gives the solution for the proportional and integral gains (kc, ki) of the controller after 73 functional evaluations. Controllers for MIMO systems can be of either multiloop (controllers are designed only for diagonal elements of process TF) or multivariable (controllers are designed for all the elements of the MIMO TF). Multiloop control[15] scheme has an edge over multivariable as the former can work even if a single loop fails. In presence of interactions between input/output, the process need to be decoupled and then multiloop controllers can be designed. Thus the aim of this article is to bring out a clear picture of interaction measure and re-design of controller for MIMO systems. The rest of the paper is carried out as follows: section 2 discusses controller parameter optimization methods of multivariable systems. Monitoring of closed loop undesired responses and redesign of the controller is explained in section 3. Performance measure is discussed in section 4. At the end, conclusion is drawn.

2. PARAMETER OPTIMIZATION

The performance of a closed-loop system is measured with a single scalar quantity-the performance index. Performance of each loop needs to be better or improved at the cost of spending optimal/minimum inputs. Thus the problem is to formulate performance criteria that will lead to find optimal solutions of manipulated inputs. A configuration of the controller is selected and free parameters of the controller that optimize the performance index need to be determined. We consider a linear 2-by-2 MIMO process and proceed to formulate performance criteria. The optimal controller parameters found from the solution will be used to retune the loops so as to minimize loss through undesirable response and unnecessary disturbance through undesired loops.

A commonly used performance criterion is the area under the regulatory response:

$$J = \int_0^\infty y_2(t)\,dt \qquad \dots (1)$$

This criterion, which has good mathematical track ability properties, is acceptable in practice as a measure of system performance. Small J results in a system with small overshoot. Since the integration is carried out over [0, ∞], a small J limits the effect of small error lasting for long time, and thus results in a small settling time. Also, a finite J implies that the steady state error is zero.

Therefore, a more realistic performance index is of the form: $J=\int_0^\infty y_2(t)\,dt$ subject to the following constraint on control signal $u(t)$, $\max|u(t)|<M$ for some constant M, M is determined by the linear range of plant.

$u^2(t)$ is a measure of the instantaneous rate of expenditure of energy.

To minimize energy expenditure, we minimize $\int_0^\infty u^2(t)\,dt$.

We would like to replace the performance criterion by the following quadratic performance index:

$$J=\int_0^\infty\left[e^2(t)+u^2(t)\right]dt \qquad \ldots (2)$$

To allow greater generality, we can insert a real positive constant λ to obtain,

$$J=\int_0^\infty\left[e^2(t)+\lambda u^2(t)\right]dt \qquad \ldots (3)$$

By adjusting the weighting factor λ, we can weight the relative importance of the system error and the expenditure of energy. By increasing λ, by giving sufficient weight to control effort, the amplitude of the control signal which minimizes the overall performance index may be kept within practical bounds, although at the expense of the increased system error.

In this paper, our intention is to show the amount of interaction obtainable in area calculation of closed loop undesirable/regulatory responses. For a general MIMO processes, a cost is minimized in changing the input control signal, u. The control pairing which needs the least area to fulfil its control targets will be the most efficient control pairing.

The design approach based on parameter optimization consists of the following steps:

Step 0: Compute the performance index (J) as a function of the free parameters k_1, k_2... k_n, of the system with fixed configuration.

$$J=J(k_1,k_2,\ \ldots,\ k_n) \qquad \ldots (4)$$

Step 1: Determine the solution set k_i of the equations,

$$\frac{\partial J}{\partial k_i} = 0; i = 1, 2...n \quad ... (5)$$

Equation (4) gives the necessary conditions for J to be minimum. From the solution set of these equations, find the subset that satisfies the sufficient conditions which require that the hessian matrix given below is positive definite.

$$H = \begin{bmatrix} \frac{\partial^2 J}{\partial {k_1}^2} & \frac{\partial^2 J}{\partial k_1 \partial k_2} & .... & \frac{\partial^2 J}{\partial k_1 \partial k_n} \\ \frac{\partial^2 J}{\partial k_2 \partial k_1} & \frac{\partial^2 J}{\partial {k_2}^2} & .... & \frac{\partial^2 J}{\partial k_2 \partial k_n} \\ . & . & . & . \\ \frac{\partial^2 J}{\partial k_n \partial k_1} & \frac{\partial^2 J}{\partial k_n \partial k_2} & .... & \frac{\partial^2 J}{\partial {k_n}^2} \end{bmatrix}$$

Since $\frac{\partial^2 J}{\partial k_i \partial k_j} = \frac{\partial^2 J}{\partial k_j \partial k_i}$, the matrix H is always symmetric.

Step 2: If there are two or more sets of k_i satisfying the necessary as well as sufficient conditions of minimization of J, then compute the corresponding J for each set. The set that gives the smallest J is optimum set.

Selection of an appropriate performance index is as much a part of the design process as the minimization of the index. In this paper, the performance of a control system can be adequately specified in terms of settling time, overshoot and steady state error. The performance index could then be chosen as,

$$J \underline{\Delta} k_1 (\text{Settlingtime}) + k_2 (\text{overshoot}) + k_3 (\text{steadystateerror}) \quad ... (6)$$

In this paper, we present the performance index which include the undesirable system characteristics and in addition have good mathematical track ability. The performance index often involves integrating closed loop regulatory response when the system is subjected a standard disturbance such as step. The system whose design minimizes the selected performance index on controller configuration is, by definition optimal.

3. CLOSED LOOP UNDESIRED RESPONSES AND REDESIGN OF CONTROLLERS

The real time experimental set-up, coupled tanks set-up is a model of a chemical plant fragment is used. Often tanks are coupled through pipes and the reactant level and flow has to be controlled. The coupled tanks experiment is designed so that the system can be configured. The coupled tanks system as shown in Figure 1 has four translucent tanks each with a pressure sensor to measure the water levels. The couplings between the tanks can be modified by the use of seven manual valves to change the dynamics of the system imposing the use of different controllers. Water is delivered to the tanks by two independently controlled, submersed pumps. Step disturbances generation is provided by four manual valves. Drain flow rates can be modified using easy to change orifice caps. The 33–041 coupled tanks version for MATLAB is controlled by using SIMULINK and an Advantech PCI 1711 interface card.

The target of the coupled tanks experiment as shown in Figure 1 is to control the level in the lower two tanks with two pumps. The process inputs are u_1 and u_2 (input voltages to the pumps) and the outputs are h_4 and h_2 (voltage from level measurement devices). The parameters of the coupled tanks experimental set up are cross-sectional area of the tanks, $A = 0.01389\ m^2$; tanks outlet area, $a = 50.2658 \times 10^{-6}\ m^2$ and gravitational constant, $g = 9.81$.

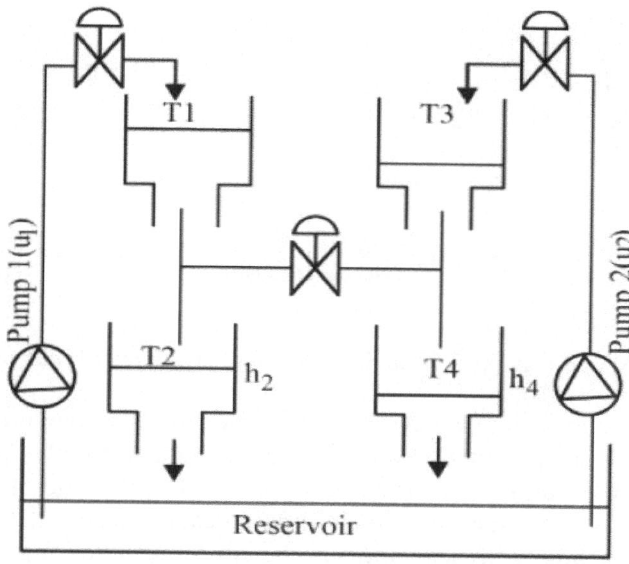

Fig. 1: Schematic Diagram of Coupled Tanks Experiment

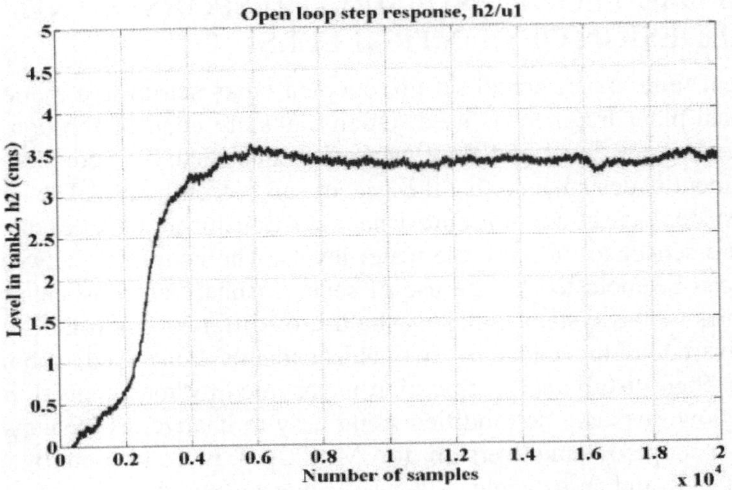

(a)

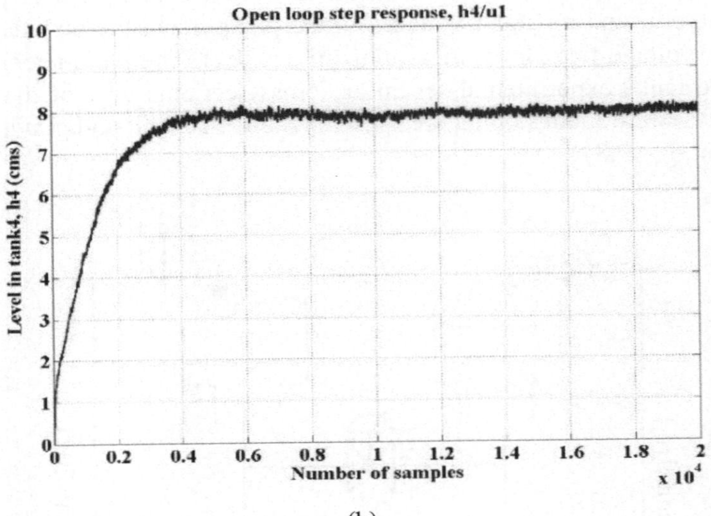

(b)

Fig. 2(a) and (b): Open Loop Step Response of Coupled Tanks System

Using the real time open loop experimental data, the process and disturbance transfer function are obtained as first order plus dead time process as in equations (7) and (8) respectively. Gain is ratio of change in level to flow rate, time constant is time taken to reach 63.2% of final steady state value and dead time is the time taken for the system to respond to its input.

$$\frac{h_4(s)}{u_1(s)} = \frac{1.0536e^{-6s}}{(279.5s+1)} \quad \ldots (7)$$

$$\frac{h_2(s)}{u_1(s)} = \frac{1.0038e^{-3s}}{(120s+1)} \quad \ldots (8)$$

Since the coupled tanks have four identical tanks, $g_{p11}(s)$ is same as $g_{p22}(s)$ and $g_{p12}(s)$ is same as $g_{p21}(s)$. The process transfer function matrix for the coupled tanks system using the open loop responses as shown in Figure 2(a) and (b) is given by Equation (9),

$$G_p(s) = \begin{pmatrix} h_2(s) \\ h_4(s) \end{pmatrix} = \begin{pmatrix} \frac{1.0038e^{-3s}}{120s+1} & \frac{1.0536e^{-6s}}{279.5s+1} \\ \frac{1.0536e^{-6s}}{279.5s+1} & \frac{1.0038e^{-3s}}{120s+1} \end{pmatrix} \begin{pmatrix} u_1(s) \\ u_2(s) \end{pmatrix} \quad \ldots (9)$$

$h_2(s)$ – Water level in tank 2 (volts)
$h_4(s)$ – Water level in tank 4 (volts)
$u_1(s)$ – Pump 1 control voltage (volts)
$u_2(s)$ – Pump 2 control voltage (volts)

The optimization toolbox routine offers a choice of algorithms. For constrained minimization, minimax, goal attainment and semi-infinite optimization, variations of sequential quadratic programming are used. Nonlinear least squares problems use the Gauss-Newton and Levenberg-Marquardt methods.

Let's say that you want to optimize the control parameters in the simulink model optsim.mdl (This model can be found in the optimization toolbox optim directory). This model includes a nonlinear process plant modeled as a simulink block diagram. The problem is to design a feedback control law that tracks a unit step input to the system. One way to solve this problem is to minimize the error between the output and the input signal. The variables are parameters of PI controllers. The routine lsqnonlin is used to perform a least squares fit. The function tracklsq must run the simulation. The simulation can be run either in the base workspace or the current workspace. To run the simulation in optsim, the variables kp, ki, a_1 and a_2 (a_1 and a_2 are variables in the plant block) must all be defined. kp and ki are the variables we are optimizing. The simulation is performed using a fixed-step fifth-order method to 100 seconds.

The controller settings and area before and after optimization are tabulated in Table 1 as follows: With these controller settings, the closed loop undesired responses obtained for coupled tanks system both before and after optimization are shown in Figure 3 shown below.

Table 1: Controller Parameters and Area under Undesired Response for Coupled Tanks System

Controller Parameters	*Before Optimization*	*After Optimization*
k_c	28.9520	41.9656
τ_I	1.4476	0.0215
Area	Before Optimization	After Optimization
	2793.2	2655.2

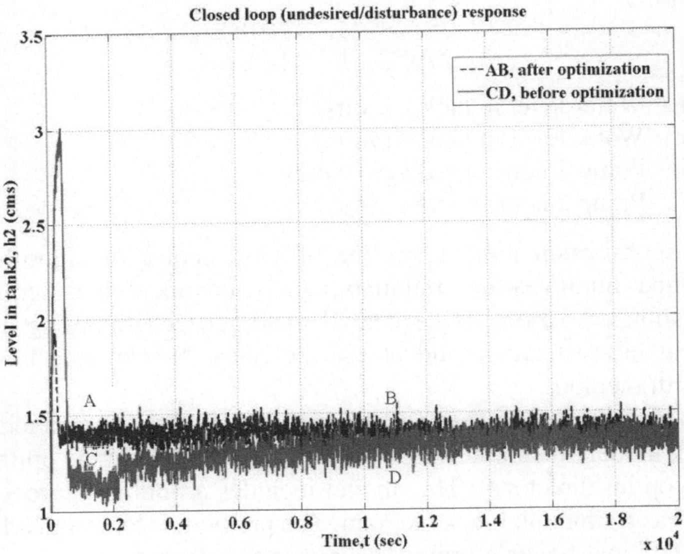

Fig. 3: Closed Loop Undesired Responses

4. PERFORMANCE MEASURE

Control effort calculated using this objective function is optimum that has saved the energy. Thus, for the coupled tanks system, gain optimum is calculated as,

$$\text{Gain Optimum} = \left(\frac{2793.2 - 2655.2}{2793.2}\right) \times 100 = 4.95\%$$

Optimum gain in this coupled tanks system is 5% when the disturbance is 10%. That is the control effort calculated using this objective function is an optimum that has saved the energy. Optimized control signal has saved 5% of utility in this particular coupled tanks system.

The performance is thus measured using standard deviation and variance and tabulated in Table 2.

Table 2: Performance Measure

Performance Measure	*Before Optimization*	*After Optimization*
Standard deviation	0.2397	0.1368
Variance	0.0574	0.0187

5. CONCLUSIONS

Thus in this paper, the proposed method of monitoring and redesign of controller have been demonstrated effectively with real time coupled tanks experimental set up. For this particular coupled tanks system, optimum gain obtained is 5% when the disturbance is 10%. Thus the control effort calculated using this objective function is an optimum that has saved the energy. A systematic approach is proposed to redesign the controller by monitoring the area under the closed loop regulatory responses. Hence, this method is designed to minimize the area under the undesired response and consequent reduction in closed loop interactions.

ACKNOWLEDGEMENTS

Authors wish to acknowledge the financial support of DST-SR/SE/CE/ 90/2009 in carrying out this research work.

REFERENCES

[1] Yu, Cheng-Ching, "Autotuning of PID controllers", Springer-Verlag London Limited, Germany, 2006.

[2] Despande, Pradeep B., "Multivariable Process control", ISA, Research Triangle Park, North Carolina, 1989.

[3] Albertos, P. and Sala, A., "Multivariable control systems: An Engineering approach", Springer-Verlag London Limited, United States of America, 2004.

[4] Bristol, E.H., "On a new measure of interaction for multivariable process control", *IEEE Trans Automatic Control*, 1966, AC-11, 133–134.

[5] Witcher, M.F. and McAvoy, T.J., "Interacting control systems: steady state and dynamic measure of interaction", *ISA transactions*, 1977, 16(3), 35–41.

[6] Tung, L. and Edgar, T.F., "Analysis of control output interactions in dynamic systems", *AIChE J.*, 1981, 27, 690–693.

[7] Lieslehto, J., "An expert system for multivariable controller design", Dr. Tech Thesis, Tampere Univ. of Tech., Finland, 1993.

[8] Saboo, A.K. and Morari, M., "Design of resilient processing plants-IV: some new results on heat exchange network synthesis", *Chemical Engineering Science,* 1984, 39(3), 579–592.

[9] Conley, A. and Salgado, M., "A Gramian based interaction measure", In: *Proceedings of 39th IEEE CDC, 2000*, Vol. 5, pp. 5020–22, Sydney, Australia.

[10] Salgado, M. and Conley, A., "MIMO interaction measure and controller structure selection", *International Journal of Control*, 2004, 77(4), 367–383.

[11] Wittenmark, B. and Salgado, M.E., "Hankel-norm based interaction measure for input-output pairing", In: *Proc of the IFAC world congress*, 2002, Barcelona, Spain.

[12] Khaki-Sedigh, Ali and Moaveni, Bijan, "Control configuration selection for multivariable plants", Springer-Verlag, Berlin, Heidelberg, Germany, 2009.

[13] Birk, W. and Medvedev, A., "A note on Grammian based interaction measures", In proceedings of European control conference, Cambridge, UK, September, 2003.

[14] Optimization toolbox for use with MATLAB, 2001, The Mathworks, Inc., 3, Apple Hill Drive, Natick, MA 01760–2098.

[15] Panda, Rames C., "Introduction to PID CONTROLLERS theory, tuning and application to frontier areas", Intech, Croatia, 2012.

...etic Algorithm Tuned ...ited Kalman Filter for Speed ...nation of Induction Motors

...a John Basha[1] and Kumar Somasundaram[2]

...trumentation Engineering, Madras Institute of Technology
...ampus, Anna University, Chennai, India
...ana.rizwan@gmail.com; [2]skumar@mitindia.edu

...ACT: *This paper aims at sensorless speed estimation of Induction ...or with minimum mean square error. The chosen algorithm for estimation is Unscented Kalman Filter. As the filter performance depends on the noise parameters such as process and measurement noise covariance which are not calculated in a straight forward approach, Genetic Algorithm is used to tune these parameters to get better performance. The result shows that the filter performance is improved using Genetic Algorithm and gives optimized minimum mean square error compared to conventional UKF under various machine operating conditions.*

Keywords: Unscented Kalman Filter (UKF), Genetic Algorithm (GA), Induction Motor.

1. INTRODUCTION

Induction Motor Drives are the most widely used Electric Drive. For Modern Induction Motor Drives sensorless control is used by which machine can be controlled without any rotational sensor. For state estimation of induction machines, nonlinear estimators are used. The rotor speed of the induction machine must be measured, which requires a speed sensor of some kind. A sensorless system where the speed is estimated instead of measured would essentially reduce the cost and complexity of the drive system. The rotor speed of the machine is not measured, and a change in load torque will result in a change in speed.

The Unscented Kalman filter (UKF) was first introduced by Julier and Uhlmann in 1997. The two widely used approximate solutions to the problem of filtering nonlinear dynamical systems are UKF and EKF. To achieve better estimation performance, the UKF has better potential than the EKF, while displaying the same order of computational complexity (3). For the UKF to work properly, it is thus necessary to tune the three scalar parameters of the unscented Transformation (UT). A common tuning strategy is given in (1) which is to make k = 3-nx, where nx is the dimension of the state vector. We call this UKF tuning a ... (UkfD). In this paper, the three scalar parameters alpha, be... are tuned using GA.

The paper is organized as follows: section 2 describes the Induction motor model used in this work, section 3 describes ... algorithm, section 4 describes the GA algorithm, section 5 descri... GA in filter tuning, section 6 describes the simulation results and an... The final conclusions are described in section 7.

2. MATHEMATICAL MODEL OF INDUCTION MOTOR

The Three phase Induction Motor dynamic model for speed sensorless control is described with two inputs and six state variables as referred in Barut *et al.* In this work the extended Induction Motor model is used where Rotor Speed is included as the fifth state variable and Torque is included as the sixth state variable.

The equation (1) shows the state space representation of Induction Motor.

The output equation is given below:

$$\begin{bmatrix} i_{s\alpha} \\ i_{s\beta} \end{bmatrix} = \begin{bmatrix} 1 & 0 & 0 & 0 & 0 & 0 \\ 0 & 1 & 0 & 0 & 0 & 0 \end{bmatrix} \begin{bmatrix} i_{s\alpha} \\ i_{s\beta} \\ \Psi_{r\alpha} \\ \Psi_{r\beta} \\ w_m \\ t_L \end{bmatrix} \quad \ldots (1)$$

The notations are defined as follows:

I_{ds}, I_{qs}	Stator current components in stationary axis
V_{ds}, V_{qs}	Stator voltage components in stationary axis
Ψ_{dr}, Ψ_{qr}	Rotor flux linkage components in stationary axis
ω_m	Rotor angular velocity
P_p	No. of pole pairs
R_s and L_s	Stator resistance and inductance, respectively
$R_s^{'}$ and $L_r^{'}$	Rotor resistance and inductance referred to stator, respectively
L_m	Mutual inductance
Σ	Leakage factor
J_L	Total inertia of the induction motor

3. UKF ALGORITHM

Unscented Kalman Filter is used to estimate mean and variance of a random variable through a known non linear function. The State equation (1) and measurement equation (2) are updated using Unscented Transformations.

Unscented Transformation

The UT can estimate the mean $\hat{y}$ and covariance P_y using mean $\hat{x}$ and covariance P_x of state variable x using 2L+1 sigma points where L is the dimension of the vector $\chi(k|k, i)$

$$\begin{bmatrix} \dot{i}_{s\alpha} \\ \dot{i}_{s\beta} \\ \dot{\psi}_{r\alpha} \\ \dot{\psi}_{r\beta} \\ \dot{w}_m \\ \dot{t}_L \end{bmatrix} = \begin{bmatrix} -\left(\frac{R_s}{L_\sigma}+\frac{R'_rL_m^2}{L_r'^2L_\sigma}\right) & 0 & \frac{R'_rL_m}{L_r'^2L_\sigma} & \frac{L_m}{L_\sigma L_r'}P_pw_m & 0 & 0 \\ 0 & -\left(\frac{R_s}{L_\sigma}+\frac{R'_rL_m^2}{L_r'^2L_\sigma}\right) & -\frac{L_m}{L_\sigma L_r'}P_pw_m & \frac{R'_rL_m}{L_r'^2L_\sigma} & 0 & 0 \\ R_r'\frac{L_m}{L_r'} & 0 & \frac{-R_r'}{L_r'} & -P_pw_m & 0 & 0 \\ 0 & R_r'\frac{L_m}{L_r'} & P_pw_m & \frac{-R_r'}{L_r'} & 0 & 0 \\ \frac{-3P_pL_m}{2J_LL_r'}\psi_{r\beta} & \frac{3P_pL_m}{2J_LL_r'}\psi_{r\alpha} & 0 & 0 & 0 & -\frac{1}{J_L} \\ 0 & 0 & 0 & 0 & 0 & 0 \end{bmatrix}$$

$$\begin{bmatrix} i_{s\alpha} \\ i_{s\beta} \\ \psi_{r\alpha} \\ \psi_{r\beta} \\ w_m \\ t_L \end{bmatrix} + \begin{bmatrix} \frac{1}{L_\sigma} & 0 \\ 0 & \frac{1}{L_\sigma} \\ 0 & 0 \\ 0 & 0 \\ 0 & 0 \\ 0 & 0 \end{bmatrix} \begin{bmatrix} v_{s\alpha} \\ v_{s\beta} \end{bmatrix} \quad \ldots (2)$$

with the associated weights W(i) are chosen symmetrically about $\hat{x}(k|k)$ as given below.

$$\chi(k|k,0) = \hat{x}(k|k) \quad \ldots (3)$$

$$\chi(k|k,i) = \hat{x}(k|k) \pm \left(\sqrt{(L+\lambda)P(k|k)}\right)_i \; i = 1, \ldots, L \quad \ldots (4)$$

$$W_m(0) = \frac{\lambda}{L+\lambda} \quad \ldots (5)$$

$$W_c(0) = \frac{\lambda}{L+\lambda} + (1 - \alpha^2 + \beta)\ ;\ \lambda = \alpha^2(L + k) - L \quad \text{... (6)}$$

$$W_c(i) = W_m(i) = \frac{1}{2(L+\lambda)}\ ;\ i = 1, \dots, 2L \quad \text{... (7)}$$

Where k is a secondary scaling parameter and can be obtained from the relation k = 3–n, where n is the dimension of the state vector.[7] Kappa is not restricted to be positive. α and β allow to capture marginal kurtosis.[6] Range of α is 1 to 10^{-4}.[5] The parameter β is used to incorporate prior knowledge of distribution of $\hat{x}$ and for Gaussian distribution its optimum value is 2.

The UKF algorithm steps is as follows:

Step 1: Initialization of State and Covariance

$$\hat{x}_0 = E[x_0]\ \text{\#Initial State} \quad \text{... (8)}$$

$$P_0 = E[(x_0 - \hat{x}_0)(x_0 - \hat{x}_0)^T\ \text{\#Initial Covariance} \quad \text{... (9)}$$

Step 2: Calculate Sigma Points,

$$\chi(k-1|k-1, i) = \hat{x}(k-1|k-1) \pm \left(\sqrt{(L+\lambda)P(k-1|k-1)}\right)_i$$

i = 1, …, L #Calculate sigma points ... (10)

Step 3: Time Update,

$$\chi(k|k-1, i) = \chi(k-1|k-1, i) + \int_{k-1}^{k} F[\chi(\tau, i), u(k-1)]d\tau;\ i = 0\dots, 2 \quad \text{... (11)}$$

$$\hat{x}(k|k-1) = \sum_{i=0}^{2L} W^m(i)\, \chi(k|k-1, i) \quad \text{... (12)}$$

#Calculate mean of the predicted state,

$$P(k|k-1) = \sum_{i=0}^{2L} W^c(i)\, \{\chi(k|k-1, i) - \hat{x}(k|k-1)\} * \{\chi(k|k-1, i) - \hat{x}(k|k-1)\}^T + Q \quad \text{... (13)}$$

#Calculate error covariance of the predicted state,

$$\chi^*(k|k-1, i) = \hat{x}(k|k-1) \pm \left(\sqrt{(L+\lambda)P(k|k-1)}\right)$$

$$i = 1, \dots, L \quad \text{... (14)}$$

#Redraw sigma points

$$\hat{y}(k|k-1) = \sum_{i=0}^{2L} W^m(i) * H[\chi^*(k|k-1, i)] \quad \text{... (15)}$$

#Calculate mean of the predicted output

Step 4: Measurement Update

$$P_{yy}(k) = \sum_{i=0}^{2L} W^c(i)\{H[\chi^*(k|k-1,i) - \hat{y}(k|k-1)\} * \{H[\chi^*(k|k-1,i) - \hat{y}(k|k-1)\}^T] + R \quad \dots (16)$$

#Calculate error covariance of the predicted output (16)

$$P_{xy}(k) = \sum_{i=0}^{2L} W^c(i)\{\chi^*(k|k-1,i) - \hat{x}(k|k-1)\} * \{H[\chi^*(k|k-1,i)] - \hat{y}(k|k-1)\}^T \quad \dots (17)$$

#Calculate Cross Covariance of state and output

$$\gamma(k) = y(k) - \hat{y}(k|k-1) \text{ \#Innovation} \quad \dots (18)$$

$$K(k) = P_{xy}(k)P_{yy}^{-1}(k) \text{ \#Kalman Gain} \quad \dots (19)$$

$$\hat{x}(k|k) = \hat{x}(k|k-1) + K(k) * \gamma(k) \quad \dots (20)$$

#Updated State estimate

$$P(k|k) = P(k|k-1) - K(k) * P_{yy}(k)K^T(k) \quad \dots (21)$$

#Updated Error Covariance

4. GENETIC ALGORITHM

Genetic Algorithm was introduced by John Holland in 1960 based on the concept of Darwin's Theory of Evolution, which is extended in 1989 by his student Goldberg.

- [Create Initial Population]: GA works on a population of some possible solutions. The population size represents the number of solutions. Each solution is called as Individual.
- [Evaluate Fitness]: Each Individual solution is represented through a chromosome(x). Each Individual has a fitness value. The best individuals are selected by fitness function f(x).
- [Evolution of New Population]: To evolve new population each iteration consists of following steps:
- [Selection]: The Selection of best individuals for reproduction is done by selecting best relative fitness of individuals. The selected Individuals are called Parents.
- [Crossover]: With crossover probability cross over the parents to form a new offspring (children). If no crossover was performed, offspring is an exact copy of parents.
- [Mutation]: With mutation probability mutate new offspring at each position in chromosome. Thus the fitness of new chromosomes are evaluated.

- [Replacement]: Individuals from old population are killed and replaced by new population.

Each population created is called a Generation.

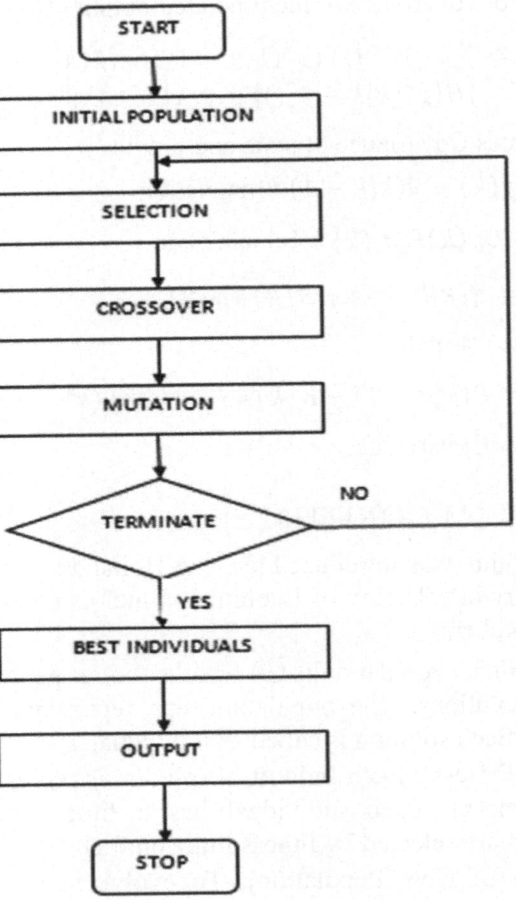

Fig. 1: Flowchart of GA

5. GA IN FILTER TUNING

To attain reliable performance of Induction Motor the UKF parameters such as alpha, beta, kappa, noise covariance matrices of rotor speed and torque are tuned using Genetic Algorithm. The objective of my work is to obtain Minimum Mean Square Error.

From Table 3, the chosen 5 parameters, Random Population is created. Fitness is evaluated for each population. For new population separate iteration loop is used. Selecting the best individuals from the population. The selected individuals are crossovered and a new offspring is created. Fitness is evaluated for new individuals. New individuals are inserted in new population. If population is converged iteration is stopped. Every new population created is assigned as generations. In my work 10 generations are created and minimum mean square error is calculated.

6. SIMULATION RESULTS AND ANALYSIS

The induction Motor specification and parameters used in this work are given in Table 4 in Appendix. In this work it is assumed that only the stator currents are measurable. Simulations have been carried out for 2000 Sampling Instants and analyzed under load change.

Speed Estimation is detected on sudden load applied to Sampling Instants and it is observed which filter estimates accurately. It is assumed that at initial load condition load is equal to zero.

At 1000th Sampling Instant Load is applied to 10 Nm. At 1500th Sampling Instant load is removed i.e. 0 Nm. The UKF parameters chosen as shown in Table 2 and GA tuned UKF are chosen as shown in Table 3.

Figure 2 shows that GA tuned UKF able to generate accurate speed estimation compared to conventional UKF.

Figure 3 shows the enlarged view of Figure 2 around 1000th Sampling Instant. Figure 4 shows the enlarged view of Figure 2 around 1500th Sampling Instant. It shows that GA tuned UKF is able to estimate more close to True value.

The Average MSE of UKF is 51.5504 and it is reduced to 3.5478 on using GA tuned UKF. The results are shown in Table 1.

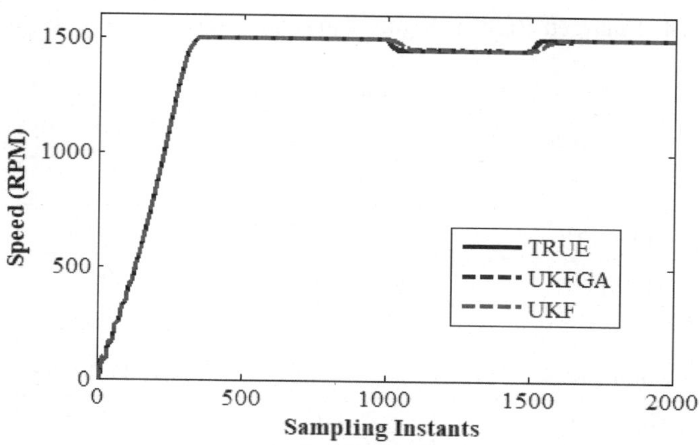

Fig. 2: Speed Estimation of the Filter

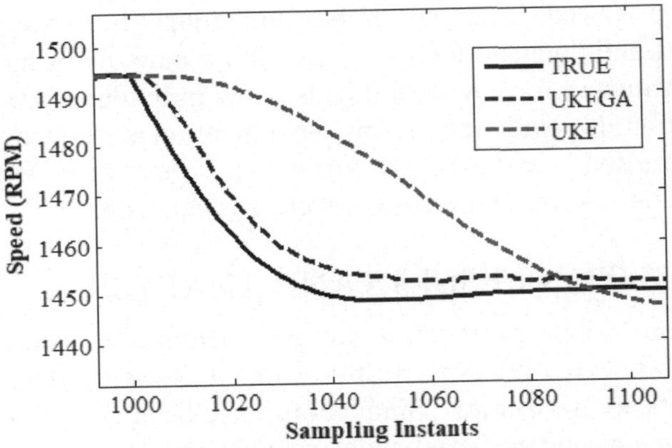

Fig. 3: Enlarged View of Figure 2 around 1000th Sampling Instant

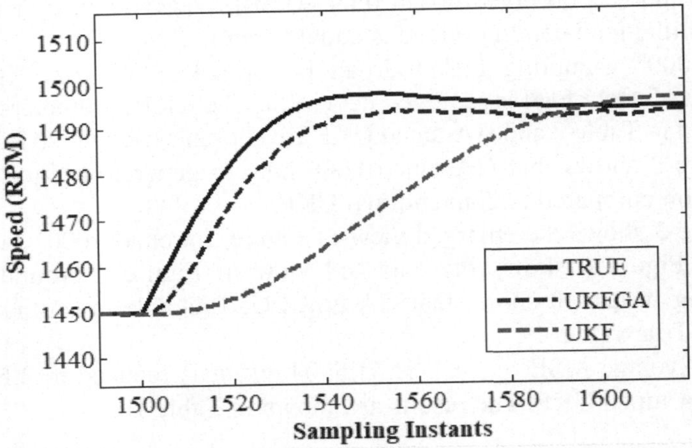

Fig. 4: Enlarged View of Figure 2 around 1500th Sampling Instant

Table 1: Average Mean Square Error for Speed Estimation

Conditions	*GA Tuned UKF*	*UKF*
Load Change		
Alpha (α)	0.9421	1.0000
Beta (β)	1.1680	2.0000
Kappa (κ)	–2.1136	–3.0000
Rotor Speed (Q5)	0.0135	0.000158
Torque (Q6)	0.0128	0.0015
Average MSE	**3.5478**	**51.5504**

Table 2: Parameters Used for UKF

Noise Covariance Matrices For UKF Algorithm			
Q1 = Q2 $5.5*10^{-3}$	Q3 = Q4 $1.5*10^{-5}$	Q5 $158*10^{-7}$	Q6 $15*10^{-4}$
Initial State Error Covariance Matrices			
P = 1000*eye(6)			
UKF Tuning Parameter Values			
$\alpha = 1$	$\beta = 2$	$k = -3$	

Table 3: Parameters Used for GA Tuned UKF

Ranges for Tuning	
α	$0.001(10^{-3})$ to 1.0
β	1.0 to 2.0
κ	–2.0 to –3.0
Q5	$158*10^{-7}$ to $158*10^{-4}$
Q6	$15*10^{-4}$ to $15*10^{-3}$
P	1000*eye(6)
Q1 = Q2	$5.5*10^{-3}$
Q3 = Q4	$1.5*10^{-5}$

Table 4: Specification and Parameters of Induction Motor

P (kW)	2.2
F (Hz)	50
J_L (kg m^2)	0.05
Poles	4
V (V)	380
I (A)	4.8
R_s (Ω)	3.3
R_r' (Ω)	2.2
L_s (H)	0.3
L_r' (H)	0.3
L_m (H)	0.286
N (rpm)	1415
T_L (Nm)	14.8

7. CONCLUSION

This paper presented to improve the performance of UKF for speed estimation of Induction Motor. The UKF performance critically depends on the proper selection of noise covariance matrices which estimates rotor speed. So GA is used to tune these parameters and select optimum value which reduces error. Simulation results shows that the GA tuned UKF estimated closer to true value than the conventional UKF method.

REFERENCES

[1] Shi, K.L., Chan, T.F., Wong, Y.K. and Ho, S.L.. "Speed estimation of an induction motor drive using extended Kalman filter." *In Power Engineering Society Winter Meeting, IEEE,* Vol. 1, pp. 243–248, 2000.

[2] Shi, S.K.L., Chan, T.F., Wong, Y.K. and Ho, S.L. "Speed Estimation of an Induction Motor Drive using an Optimized Extended Kalman Filter", *IEEE Transactions on Industrial Electronics,* Vol. 49, Issue 1, pp. 124–133, 2002.

[3] Barut, Murat; Bogosyan, Seta and Gokasan, Metin. "Speed-sensorless estimation for induction motors using extended Kalman filters." *IEEE Transactions on Industrial Electronics*, 54, No. 1, pp. 272–280, 2007.

[4] Julier, Simon J., Uhlmann, Jeffrey K. and Durrant-Whyte, Hugh F. "A new approach for filtering nonlinear systems." *In American Control Conference, Proceedings of the 1995,* Vol. 3, pp. 1628–1632. IEEE, 1995.

[5] Wan, Eric A. and Merwe, Rudolph Van Der. "The unscented Kalman filter for nonlinear estimation." *In Adaptive Systems for Signal Processing, Communications, and Control Symposium 2000. AS-SPCC IEEE*, pp. 153–158, 2000.

[6] Ponomareva, Ksenia and Date, Paresh. "Higher order sigma point filter: A new heuristic for nonlinear time series filtering." *Applied Mathematics and Computation 221,* pp. 662–671, 2013.

[7] Julier, Simon J. and Uhlmann, Jeffrey K. "Reduced sigma point filters for the propagation of means and covariances through nonlinear transformations." In American Control Conference, 2002. *Proceedings of the 2002,* Vol. 2, pp. 887–892. IEEE, 2002.

[8] Akin, Bilal; Orguner, Umut and Ersak, Aydin. "State estimation of induction motor using unscented Kalman filter." In Control Applications, 2003. CCA 2003. *Proceedings of 2003 IEEE Conference on*, Vol. 2, pp. 915–919, 2003.

[9] Laamari, Yahia; Chafaa, Kheireddine and Athamena, Belkacem. "Particle swarm optimization of an extended Kalman filter for speed and rotor flux estimation of an induction motor drive." *Electrical Engineering 97,* No. 2 pp. 129–138, 2015.

[10] Lin, Guohan, Zhang Jing and Liu, Zhaohua. "Tuning of extended Kalman filter using improved particle swarm optimization for sensorless control of induction motor." *Journal of Computational Information Systems 10,* No. 6 pp. 2455–462, 2014.

[11] Kumar, S., Prakash, J. and Kanagasabapathy, P. "A critical evaluation and experimental verification of Extended Kalman Filter, Unscented Kalman Filter and Neural State Filter for state estimation of three phase induction motor." *Applied Soft Computing 11*, No. 3 pp. 3199–3208, 2011.

[12] Horváth, K. and Kuslits, M., "Optimization-based parameter tuning of unscented Kalman filter for speed sensorless state estimation of induction machines," *2017 5th International Symposium on Electrical and Electronics Engineering (ISEEE),* Galati, pp. 1–7, 2017.

[13] Chand, B.K., Khashirunnisa, S. and Kumari, B.L., "Performance analysis of Kalman filter with Particle Swarm Optimization and fuzzy logic," *2016 International Conference on Electrical, Electronics, and Optimization Techniques (ICEEOT),* Chennai, pp. 3204–3208, 2016.

[14] Goldberg, D.E. (1989). Genetic Algorithms in Search, Optimization, and Machine Learning. Reading: Addison-Wesley.

Optical based Instrumentation Setup to Detect Potassium under High Pressure Condition

J. Sumathi[1] and V. Sathiesh Kumar[2]

Department of Electronics Engineering, MIT Campus, Anna University, Chennai

E-mail: [1]sumathi.j05@gmail.com; [2]sathiieesh@gmail.com

ABSTRACT: *Potassium identification under high pressure condition (1–8 atm) is carried out using optical based instrumentation. The technique used for measurement is Laser Induced Breakdown Spectroscopy technique (LIBS). The parameters such as laser wavelength (1064 nm, 532 nm, 355 nm), laser fluence (1–3 J/cm^2) and ambient pressure (1–8 atm) is studied for possible detection of potassium from potassium iodide (KI) salt. Plasma temperature is also estimated. From the experimental analysis, it is shown that the optical based instrumentation setup is suitable to determine the elemental composition (Potassium) under high pressure condition.*

Keywords: Spectral Measurement, Laser Induced Breakdown Spectroscopy, Plasma Temperature.

1. INTRODUCTION

Potassium (K) is one of the important minerals responsible for human body's biochemical function. In India, the land based potash resources are very limited. The ocean environment are rich in potassium content. But the identification and retrieval of potassium under high pressure condition is a challenging task. As the depth of the ocean bed increases, the ambient pressure also gets increased thereby the identification of potassium is difficult.[1]

In this paper, Laser Induced Breakdown Spectroscopy (LIBS) technique is used to identify the potassium content under high pressure condition (1–8 atm). This technique is currently used in various fields irrespective of its state (solid, liquid, gas) for both qualitative and quantitative analysis. This technique is a multi-element detection technique and does not require sample preparation. It is an in-situ technique and can be used in any harsh environment such as ocean environment.[1] Y. Dixit *et al*. quantified the elemental composition of minced beef using LIBS combined with chemometrics (Partial Least Square Regression-PLSR) as a monitoring tool. Authors have also detected the adulteration in beef kidney. Atomic absorption spectroscopy (AAS) is used to determine the reference values of sodium (Na) and potassium (K). Based on the prediction percentage of Na and K spatial mineral maps of minced beef samples are created. For all the models good prediction accuracy is observed.[2]

Y. Groisman *et al.* tested the feasibility of Laser Induced Breakdown Spectroscopy technique (LIBS) for online fertilizers production control. Authors have reported that it is possible to analyze all the relevant elements such as K, Na and Mg.[3] Maria Markiewicz-Keszycka *et al.* tested the potential of LIBS combined with chemometric to monitor the ash, potassium and magnesium content in gluten free flours (tapioca, potato, maize, buckwheat, brown rice and gluten free powder mixture). Principal component analysis is used to differentiate different flours. Partial least square regression is used for quantification.[4]

Da Chen *et al.* carried out the real time quantification of potassium content in infant formula using Data Driven LIBS (DD-LIBS). To select the most important features related to potassium accurately and without any interference Modified Random Frog Algorithm (MRFA) is included in higher-density discrete wavelet transform (HDWT). Reconstructed spectra using DD-LIBS are highly consistent with the real spectra. Authors have also reported that quantification of any solid samples with uncontrolled interference could be performed using DD-LIBS.[5]

Based on the literature survey it has been identified that LIBS based instrument could be used for identification of potassium content in samples. It is also identified that the identification of potassium under high pressure conditions are very spares. In this paper, optical based instrumentation method is used to identify the potassium content under high pressure condition. The influence of incident laser wavelength and fluence is also studied in terms of potassium identification under high pressure condition.

2. EXPERIMENTAL SETUP

Figure 1 shows the experimental setup used for the identification of potassium in samples under high pressure condition. Q-switched Nd: YAG (Quantel-Q smart 450) laser is used to ablate the material under analysis. It has a fundamental wavelength of 1064 nm, repetition rate of 10 Hz and a pulse duration of 5 ns. High energy laser beam is focused on the target using a planoconvex lens of focal length 300 mm and diameter 50 mm. The sample is placed inside a high pressure chamber on a rotating target, such that fresh samples get ablated every time. The self-designed stainless steel cylindrical high pressure chamber can withstand a maximum of

8 atm. It has 6 quartz windows to collect optical emission. 4 on the circumference of the chamber, 1 at the top and 1 at an angle of 45°. Samples are made as pellets. 1 gm of potassium iodide salt (KI) is placed on a die set, pressure (2 Ton) is applied using a manual hydraulic press for 2 minutes. Due to ablation plasma is created. The optical emission from the plasma is focused into fiber optic cable using collecting lens of focal length 100 mm and diameter 50 mm. To reduce the continuum radiation spectrometer is triggered after 6 μs from the laser pulse using delay generator (Model No: DG645 Digital Delay/Pulse Generator). The optical emission collected from plasma is spectrally studied using spectrometer (Ocean optics 2000+, Resolution: 1 nm).

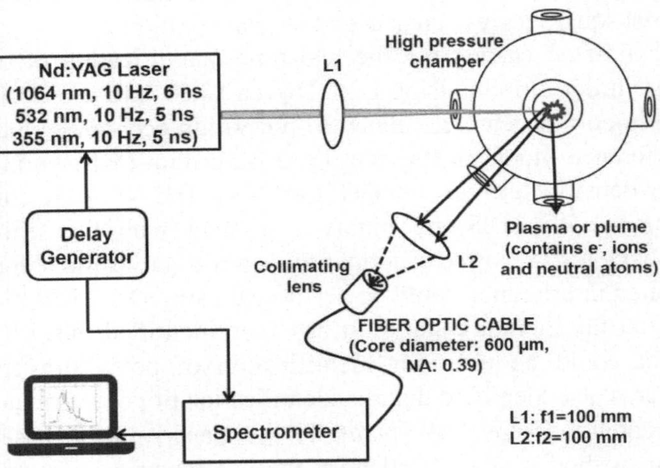

Fig. 1: LIBS Experimental Setup

3. RESULTS AND DISCUSSION

Optical based instrumentation is used to detect the presence of potassium under high pressure condition (1–8 atm). Spectral analysis is performed by varying the laser fluence (1–3 J/cm^2), laser wavelength (1064 nm, 532 nm, 355 nm) and ambient pressure (1–8 atm).

3.1 Spectral Measurement

Figure 2 (a–c) shows the spectra of KI salt for varying laser wavelength and fluences. It is observed that the optical emission lines of neutral potassium (KI) are visible at 404.414 nm, 691.108 nm, 693. 628 nm, 766.489 nm and 769.896 nm. Singly ionized and doubly ionized lines are not visible due to the requirement of high ionization energy. From the graph, it is observed that as the laser fluence increases, the measured intensity of neutral atoms also increases. It is also observed that the neutral atom emission lines are visible irrespective of incident laser wavelength (1064 nm, 532 nm, 355 nm). Emission lines due to iodine are not observed due to the requirement of high ionization energy (10.451 eV).[6] This is due to the fact that with an increase in the laser fluence the amount of material ablated gets increases which in turn increases the emission line intensity.

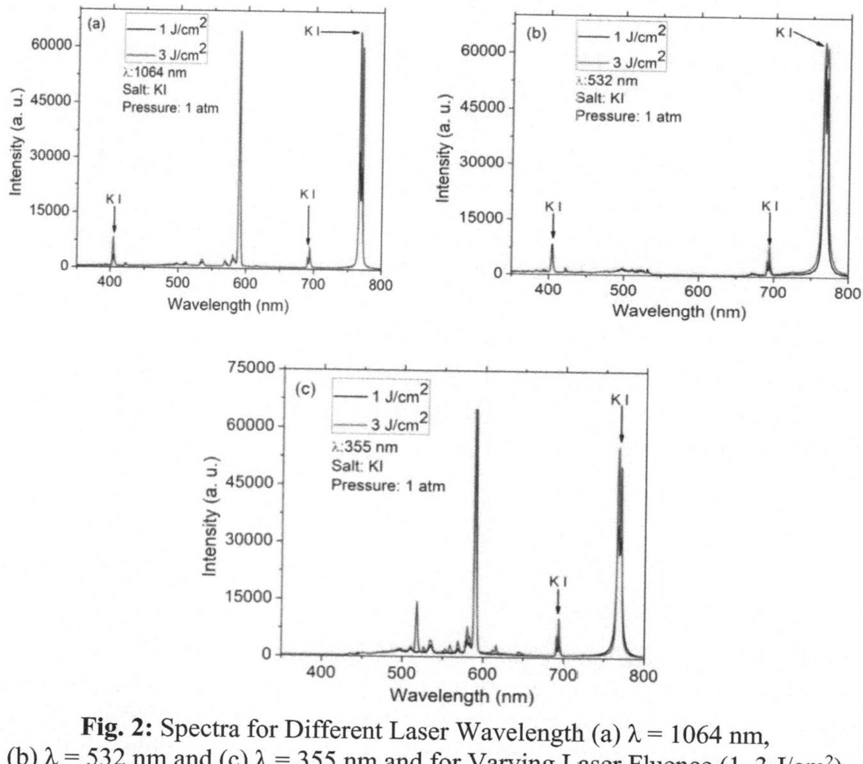

Fig. 2: Spectra for Different Laser Wavelength (a) λ = 1064 nm, (b) λ = 532 nm and (c) λ = 355 nm and for Varying Laser Fluence (1–3 J/cm^2)

3.2 Spectral Analysis for Varying Ambient Pressure

Similar spectral analysis is carried out for varying ambient pressure condition (1–8 atm) for a constant laser wavelength of 1064 nm and for varying laser fluence (1–3 J/cm^2), Figure 3 shows the spectra for varying pressure (1–8 atm) and at a laser wavelength of 1064 nm for (a) 1 J/cm^2 and (b) 3 J/cm^2.

From the graph, it is observed that emission line intensity drops with a raise in the ambient pressure for constant laser fluence. This might be due to the fact that the ambient pressure does not allow the plasma to expand beyond a limit thereby results in plasma shielding. Huaming Hou *et al.* observed similar trends in their studies for varying pressure conditions (1–395 atm).[7] It is also observed that for a constant pressure the emission line intensity increases with the increase in the laser fluence.

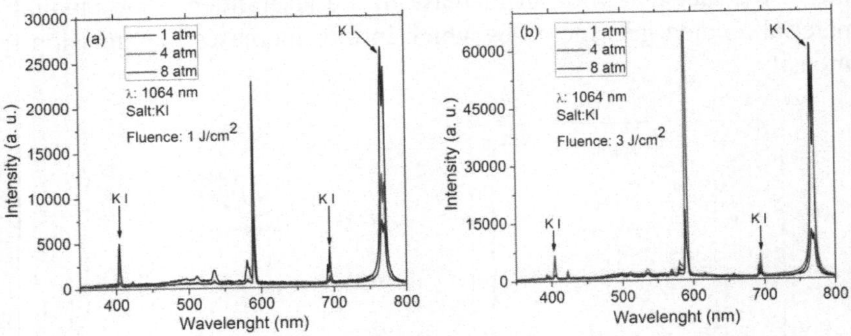

Fig. 3: Spectra Obtained for Different Laser Fluence (a) 1 J/cm^2 and (b) 3 J/cm^2 for Varying Ambient Pressure (1–8 atm)

3.3 Estimation of Plasma Temperature

Assuming Local Thermodynamic Equilibrium, the plasma parameter such as plasma temperature is estimated using line pair method using the following equation,[8]

$$T_e = 1.44\left(\frac{(E_2 - E_1)}{\ln\left(\frac{I_1\lambda_1 A_2 g_2}{I_2\lambda_2 A_1 g_1}\right)}\right) \quad \ldots (1)$$

where, T_e is the plasma temperature (K), I_1 and I_2 are the intensity of two emission lines of same species at λ_1 and λ_2, respectively, λ_1 and λ_2 are two emission wavelength of same species, E_1 and E_2 are the energies of two emission lines at λ_1 and λ_2, respectively. A_1, A_2 transition probability at λ_1, λ_2, respectively g_1, g_2 are statistical weight for the upper level at λ_1, λ_2, respectively. Parameters used for the estimation of plasma temperature is listed in Table 1.[6]

Figure 4(a) and (b) shows the plasma temperature estimated for varying laser wavelength and ambient pressure respectively. From the Figure 4(a), it is observed that for different laser wavelengths the plasma temperature does not vary significantly with the varying laser fluence. From Figure 4(b), for a constant laser wavelength and for increase in the ambient pressure and laser fluence, the plasma temperature increases. This is due to plasma condensation effect during the evolution of plasma, which in turn increases the plasma temperature. Similar trends were observed by Huaming Hou *et al.*[7]

Table 1: Spectral Parameters Used for Plasma Temperature Estimation[6]

Line	*Wavelength λ (nm)*	*Energy E_k (cm^{-1})*	*$A_k g_k$ (s^{-1})*
KI	691.108	27450.7104	$5.44e^6$
	693.628	27398.147	$1e^5$
	766.489	13042.8960	$1.52e^8$
	769.896	12985.1857	$7.50e^7$

From the experimental analysis, it is concluded that the optical based instrumentation proposed is suitable for identification of potassium content under high pressure condition.

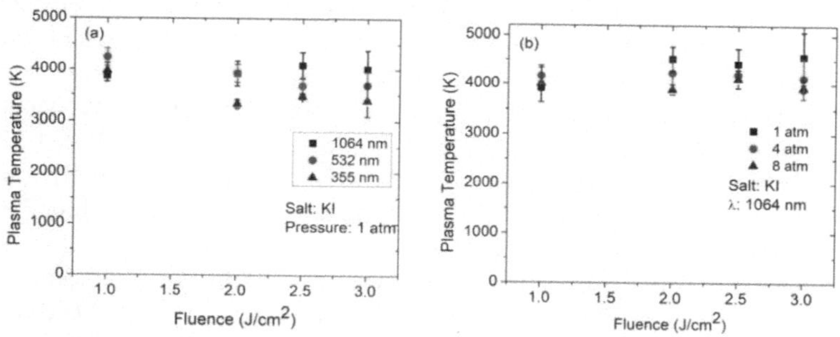

Fig. 4: Plasma Temperature for (a) Varying Laser Wavelength and (b) Varying Ambient Pressure

4. CONCLUSION

In this paper, Laser Induced Breakdown Spectroscopy (LIBS) technique is used to identify the potassium content under high pressure conditions. The experiments are carried out by varying laser wavelength, fluence and ambient pressure. It is observed that the KI lines at (404.414 nm, 691.108 nm, 693. 628 nm, 766.489 nm and 769.896 nm) are observed for different incident laser wavelength. It is also observed the emission lines due to singly ionized and doubly ionized atoms are not visible due to the requirement of large ionization energies. Emission lines due to iodine are also not observed. As the laser fluence is increased, the emission line intensity also gets increased. As the pressure increases, the emission intensity gets decreased. The plasma temperature is estimated considering LTE condition. It is observed that the plasma temperature is of about approximately 4000 K ± 500 K.

ACKNOWLEDGEMENT

This work is supported by the grant-in-aid from the Science and Engineering Research Board-File no ECR/2015/000028 dated 10.08.2016.

REFERENCES

[1] Hou, H., Tian, Y., Li, Y. and Zheng, R., "Study of pressure effects on laser induced plasma in bulk seawater", *Journal of Analytical Atomic Spectrometry*, Vol. 29(1), pp. 169–175, 2014.

[2] Dixit, Y., Casado-Gavalda, Maria P., Cama-Moncunill, R., Cama-Monucunil, X., Markiewicz-Keszycka, Maria, Cullen, P.J. and Sullivan, Carl, "Laser induced breakdown spectroscopy for quantification of sodium and potassium in minced beef: A potential technique for detecting beef kidney adulteration," *Analytical Methods*, Vol. 9(22), pp. 3314–3322, 2017.

[3] Groisman, Y. and Gaft, M., "Online analysis of potassium fertilizers by Laser-Induced breakdown spectroscopy", *Spectrochimica Acta Part B: Atomic Spectroscopy*, Vol. 65(8), pp. 744–749, 2010.

[4] Markiewicz-Keszycka, Maria; Casado-Gavalda, Maria P.; Cama-Moncunill, Xavier; Cama-Moncunill, Raquel; Dixit, Yash; Cullen, Patrick J. and Sullivan, Carl, "Laser-Induced breakdwon spectroscopy (LIBS) for rapid analysis of ash, potassium and magnesium in gluten free flours", *Food chemistry*, Vol. 244, pp. 324–330, 2018.

[5] Chen, Da; Zong, Jing; Huang, Zhixuan; Liu, Junxin and Li, Qifeng, "Real-Time analysis of potassiu in infant formula powder by Data-Driven Laser-Induced breaksown Spectroscopy", *Frontiers in chemistry*, Vol. 6, pp. 325, 2018.

[6] NIST Handbook of basic atomic spectroscopic data. http://physics.nist.gov/PhysRefData/Handbook/index.htmls.

[7] Hou, Huaming; Yu, Zenghui and Zheng, Ronger, "Plasma condensation effect induced by ambient pressure in laser-induced breakdown spectroscopy", *Applied Physics Express*, Vol. 7(3), pp. 032402, 2014.

[8] Descoeudres, A., Hollenstein, Ch., Demellayer, R. and Walder, G., Optical emission spectroscopy of electrical discharge machining plasma. *J. Phys. D Appl. Phys.*, 37, 87582, 2004.

Automatic Recognition of Pests in Greenhouse Using Image Processing Techniques

C. Shanthi* and B. Revathy

...rumentation Engineering, MIT Campus, Anna University

*E-mail: cgshanthi@gmail.com

...lant pest recognition and detection is vital in food security, ... life and stable agricultural economy. An automatic pest ...ation system using image processing technique is reported in this ...ct. The image segmentation technique is used to detect the presence ... different types of pest in leaf images. The performance of clustering-based image segmentation depends on image simplification. The K-means cluster algorithm has been used to identify the accurate location of whiteflies, aphids, thrips, dipetra and spider mites in the leaf. A multi-class Support Vector Machine (SVM) is used to classify the different types of pest present on greenhouse based on trained features. The pest infected area is also calculated by SVM classifier. Morphological operations are used to remove the unwanted elements in the classified image. This algorithm has achieved the ideal operating speed and precision. It provides decision support and also has good application prospect in agriculture.

1. INTRODUCTION

Agriculture is heart of our civilization. But Plant diseases have turned into a dilemma as it can cause significant reduction in both quality and quantity of agricultural products. Nowadays farmers are facing many crucial problems for getting better yield cause of rapid change in climate and unexpected level of insects, in order to get better yield, need to reduce the level of pest insect. India is an agricultural country wherein about seventy percentage of the population depends on agriculture. Farmers have wide range of diversity to select suitable fruit and Vegetable crops. However, the cultivation of these crops for optimum yield and quality product is highly technical. It can be improved with the aid of technological support. The management of perennial fruit crops requires close monitoring especially for the management of diseases that can affect production significantly and subsequently the post-harvest life. In case of plant the disease is defined as any impairment of norm al physi... function of plants, producing characteristic symptoms.

2. LITERATURE REVIEW

Several studies are reported in literature for detection and cla of leaf pests.

In the year 2013, H. Al-Hiary, S. Bani-Ahmad, M. Reyalat perf work "Fast and accurate detection and classification of plant dis Here, automatic detection and classification of the plant diseases have implemented using k-means segmentation algorithm followed by fea extraction using colour co-occurrence method and classification us artificial neural networks.

In the year 2011, Y. Tian, L. Wang and Q. Zhou performed a work "grading method of crop disease based on image processing". In this paper, a technique has been proposed where expanse and grading time is reduced as compared to manual system.

3. PROPOSED METHODOLOGY

In our proposed model the k-mean clustering is applied sequentially. Initially on the image was clustered using k-mean clustering. The algorithm works iteratively to assign each data point to one of K groups based on the features that are provided. K-Means clustering is a fast, robust,

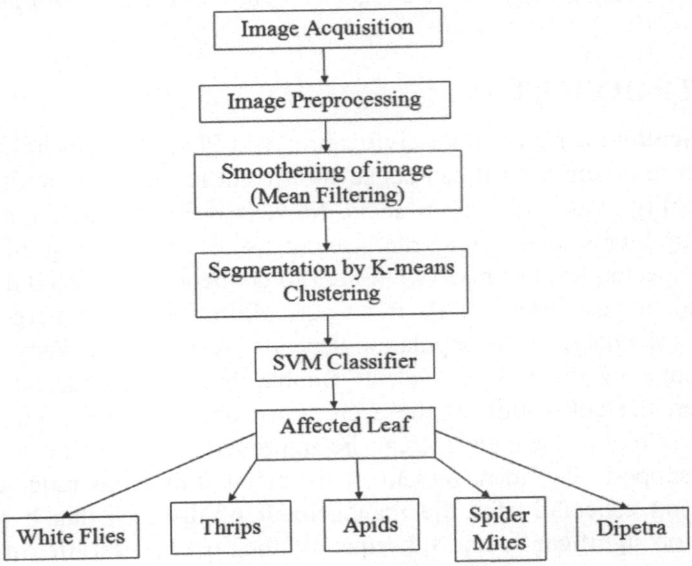

Fig. 1: Flowchart of the Proposed Method

and simple algorithm that gives reliable results when data sets are distinct or well separated from each other in a linear fashion. It is best used when the number of cluster centers, is specified due to a well-defined types. A Support vector machine is a powerful tool for binary classification, capable of generating very fast classifier function following a training period. There are several approaches to adopting SVMs to classification problems with three or more classes.

The architecture of proposed system consists various stages, including a collection of images of pests for creation of the database. Clustering techniques used for image segmentation. Features of segmented images are stored in a database with respective image of agricultural pests. The proposed approach is implemented in MATLAB.

3.1 Image Acquistion

The first step of every image processing application is image acquisition or image capturing. The images of leaves are captured by using the camera and it will store it in some formats like .PNG, .JPG, .JPEG etc.

3.2 Image Pre-Processing

Pre-processing is process which helps us to reduce the background noise. Intensity of the image should be normalized. By Enhancing input image or image captured by digital camera, is to remove the background noise, image can get enhanced visual appearance of input images. The image pre-processing steps used in the system are:

1. Conversion of RGB image to grey image
2. Resizing of the image
3. Filtering of the image.

3.2.1 *Conversion of RGB to Gray Image*

In RGB colour model, each colour appears in its primary spectral components of red, green, and blue. The disadvantages of RGB models are, it requires large space to store and it will take more time to process. So there is a need for converting the RGB model to Grey model.

3.2.2 *Resizing of the Image*

Resizing is an important step in image pre-processing. The acquired image is resized according to the requirement of the system. Resizing is nothing but, changing the dimensions of an image. The captured image is resized using some resizing methods according to the requirement of the system.

3.2.3 *Filtering*

This is also a process which comes under pre-processing a mean filtering technique is used to remove unwanted noise in enhanced image. The median filtering technique is used by mean to calculate PSNR value.

Fig. 2: Whiteflies

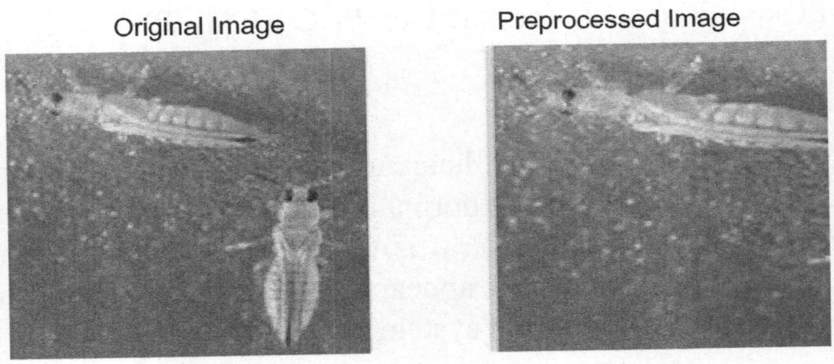

Fig. 3: Thrips

4. SEGMENTATION

Image segmentation is the process of partitioning a digital image into multiple segments. The result of image segmentation is a set of segments that collectively cover the entire image, or a set of contours extracted from the image. Each of the pixels in a region is similar with respect to some characteristic or computed property, such as colour, intensity, or texture.

4.1 K-Means Clustering

K-means clustering is a type of unsupervised learning, which is used when you have unlabelled data (i.e., data without defined categories or groups). The goal of this algorithm is to find groups in the data, with the number of groups represented by the variable *K*. The algorithm works iteratively to assign each data point to one of *K* groups based on the features that are provided. Data points are clustered based on feature similarity. The results of the *K*-means clustering algorithm are:

1. The centroids of the K clusters, which can be used to label new data.
2. Labels for the training data (each data point is assigned to a single cluster).

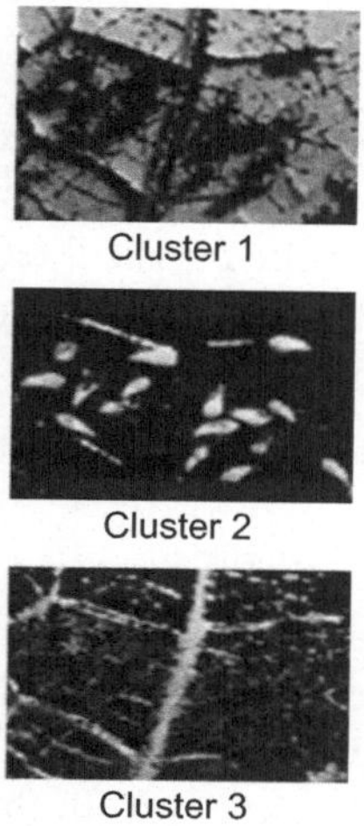

Fig. 4: Segmented Image of Dipetra

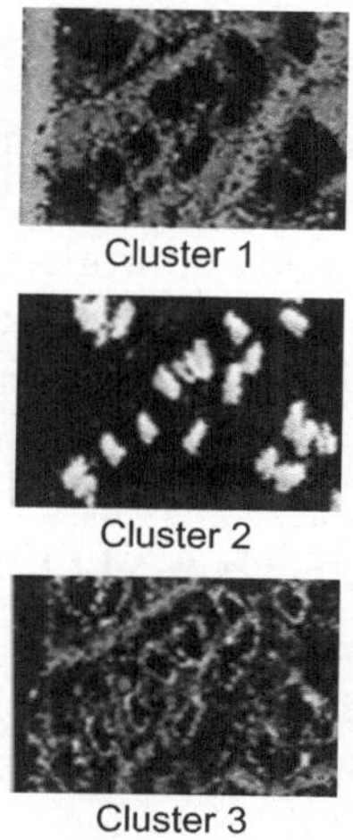

Fig. 5: Segmented Image of Whiteflies

5. FEATURE EXTRACTION

Feature extraction is the most important part of this project. The properties standard deviation, entropy, contrast etc. are extracted from the image and are used to train the dataset for the classification. Image processing is the analysis and manipulation of graphical images from sources such as photographs and videos. The most important part of this segmentation method is extension of feature space. But in real images, noise is corrupting the image data or image usually consists of textured segments. This phase is the extraction of the detected insect pest from the image. The output image which was obtained at the end of the previous phase was used in this phase. The image pixel values of the output image will be scanned both horizontally and vertically to determine the coordinates of each insect in the image. The width and height of the extracted image was determined by using its starting and ending coordinates.

6. CLASSIFICATION

Image Segmentation is the classification of an image into different groups. SVM is a supervised machine learning algorithm which can be used for classification or regression problems. It uses a technique called the kernel trick to transform your data and then based on these transformations it finds an optimal boundary between the possible outputs. SVM it capable of doing both classification and regression. The benefit is that you can capture much more complex relationships between your data points without having to perform difficult transformations on your own.

6.1 Multi-SVM

Multi class SVMs are a new technique suitable for classification tasks, which is related to and contains elements of non-parametric applied statistics, neural networks and machine learning. Like classical techniques, multi SVMs also classify a company as solvent or insolvent according to its score value, which is a function of selected financial ratios. But this function is neither linear nor parametric.

7. RESULTS AND DISCUSSION

All the image pre-processing, segmentation, feature extraction and classification techniques in our proposed method are stimulated in MATLAB. After pre-processing and segmentation steps, 10 texture features are extracted from these images as described in feature extraction. In this project work, the experiments are carried using MATLAB. I have prepared a GUI layout with a list of menus. Clicking on each menu will perform an independent function. The processing scheme consists of test RGB image acquisition from database.

Image pre-processing includes image enhancement and image segmentation where the images are segmented by K-means Clustering and classified by means of multiclass SVM Classifier. The input image is collected using digital camera. The captured image is pre-processed. The pre-processed image and the image after classifying the pest infected region by using k-means colour based segmentation technique are given. Finally, the accuracy of infected region is also calculated by SVM classifier.

Table 1: Types of Pest and Infected Area

Pest Name	Cluster 1	Cluster 2	Cluster 3
White flies	15.0031%	16.0045%	17.5634%
Thrips	16.1588%	16.6745%	15.4540%
Aphids	12.3457%	12.3346%	14.4576%
Spider mites	13.3443%	14.3378%	15.6657%
Dipetra	16.7489%	15.3356%	170.061%

Table 2: Extracted Features for 5 Different Types of Pests

Parameter	White Flies	Thrips	Aphids	Spider Mites	Diptera
Mean	24.857	25.098	24.526	19.2565	18.5625
Standard Deviation	65.1178	59.2569	59.5623	57.568	56.2235
Entropy	1.92977	1.4563	4.5628	5.2689	5.2269
RMS	5.92785	5.0096	8.9463	5.2368	5.0026
Variance	3972.57	3987.66	3480.7	4582.22	3986.22
Kurtosis	8.01108	3.2568	3.2356	2.3356	2.2235
Skeewness	2.5477	0.6166	1.2962	0.7112	0.6145
Contrast	0.619332	0.91148	0.296032	0.97785	0.91175
Energy	0.699636	0.6972	0.77123	0.7125	0.69485
Homogeneity	0.0942531	0.942001	0.883	0.88001	0.97586

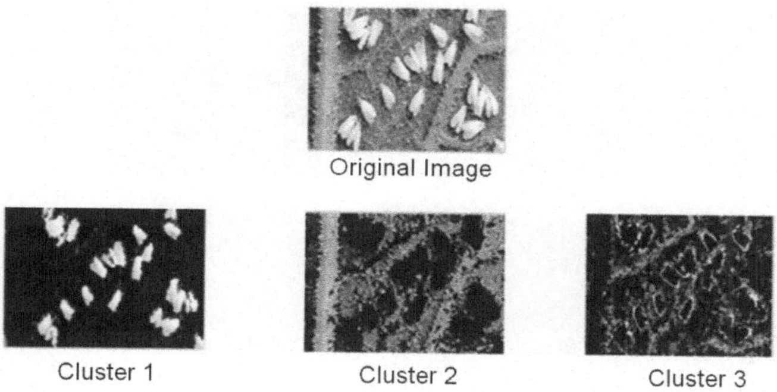

Fig. 6: Selected Clear Pest Image form Classification Result of Image 1

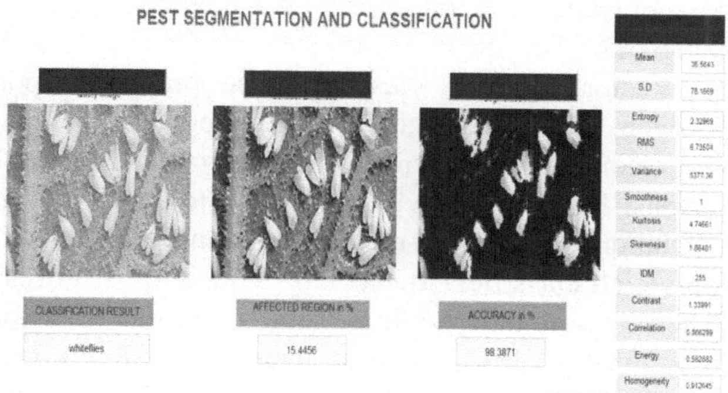

Fig. 7: Classification Result for the Pest Infected Leaf Image 1

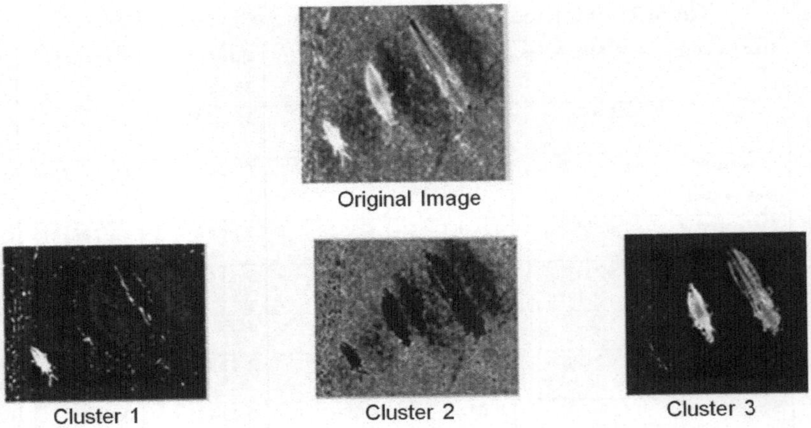

Fig. 8: Selected Clear Pest Image form Classification Result of Image 4

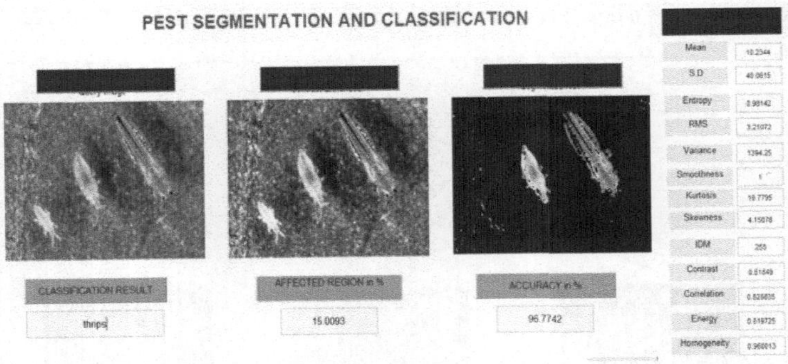

Fig. 9: Classification Result for the Pest Infected Leaf Image 4

8. CONCLUSION

Image processing techniques for several plant species that have been used for recognizing pest. The major techniques used are K-means clustering and Support Vector Machine classifier. The proposed approach is a valuable approach, which can significantly support an accurate detection of pest in a little computational effort. The system proved reliable for rapid detection of pests. This work provides a method to detect pest using image processing technique. Firstly, K-means clustering algorithm is used for segmentation which classifies objects based on set of features into K no. of classes where feature extraction is feature variance used for matching the train image features from database images and finally recognition is performed using SVM Classifier. Our goal is to detect the pests as early as possible and reduce the use of pesticides.

REFERENCES

[1] Haysshi, S., Ganno, K., Ishii, Y., *et al.* (2012). "Robotic harvesting system for eggplants," *Japan Agricultural Research Quarterly*, Vol. 36, pp. 163–168.

[2] Takahashi, T., Zhang, S., Fukuchi, H., *et al.* (2014). "Binocular stereo vision system for measuring distance of apples in orchard (Part 1)-Method due to composition of left and right images," *Journal of the Japanese Society of Agricultural Machinery*, Vol. 62, pp. 89–99.

[3] Takahashi, T., Zhang, S., Fukuchi, H., *et al.* (2013). "Binocular stereo vision system for measuring distance of apples in orchard (Part 2)-Analysis of and solution to the correspondence problem," *Journal of the Japanese Society of Agricultural Machinery,* Vol. 62, pp. 94–102.

[4] Kondo, N., Ninomiya, K., Hayashi, S, *et al.* (2015). "A new challenge of robot for harvesting strawberry grown on table top culture," ASAE paper, No. 021102, ASAE.

[5] Hannan, M.W., Burks, T.F. and Bulanon, D.M. (2009). "A real-time machine vision algorithm for robotic citrus harvesting," ASAE paper No. 073125, ASAE.

[6] Kimiko, Gotou; Tateshi, Fujiura; Yoshifumi, Nishiura and Hideo, Ikeda (2014). "3-D vision system of tomato production robot," *International Conference on Advanced Intelligent Mechatronics.*

[7] Clement, A., Verfaille, T., Lormel, C. and Jaloux, B. (2015). "A new colour vision system to quantify automatically foliar discolouration caused by insect pests feeding on leaf cells". *Biosys. Eng.* 133, 128–1.

[8] Gonzalez, R.C. and Woods, R.E. (2012). "Digital Image Processing. Prentice Hall Upper Saddle River", New Jersey 07458, 797, pp. 80–81.

Design of Regression Model to Monitor Reheater Ash Fouling

S. Anitha Kumari[1] and S. Srinivasan[2]

[1]Department of Biomedical Engineering, Saveetha School of Engineering, Chennai

[2]Department of Instrumentation Engineering, Madras Institute of Technology Campus, Anna University, Chennai

E-mail: [1]anithalakshmi_s@yahoo.com; [2]srini@mitindia.edu

ABSTRACT: *Coal ash deposition represents a major operational challenge as it entails significant losses in the heat transfer efficiency. Soot blowers are activated on a predefined cleanliness level in order to maintain the required level of heat transfer for the current operating conditions. In this paper ARIMAX model is proposed to predict the cleanliness of reheater that could aid in targeted approach of soot blow operation.*

Keywords: Cleanliness Factor, Ash Fouling, ARIMAX, Reheater.

1. INTRODUCTION

Ash foul monitoring is critical to ensure efficient operation. Cleanliness Factor (CF) is a widely used performance parameter to retrieve the extent of fouling.[6] It is defined as the ratio between the overall heat transfer coefficient of fouled and clean heat transfer surfaces. To permit targeted approach of soot blow operation based on boiler operating conditions estimation of cleanliness factor is essential. Ash fouling of economizer tubes has been detected using a physical model that incorporates the effect of tube metal temperatures.[2] Polynomial fuzzy observer[7] has been developed for fouling detection at both transient and steady state conditions. Threshold fouling model were employed to determine fouling quantitatively in terms of deposition and inhibition mechanisms.[8] The reliability of the threshold models were fair with laboratory set up but fail when applied to industrial situations due to the variability of fouling behaviour under the influence of several process parameters. Fouling detection is achieved by using CF as output of static neural network model with mass flow rate and outlet temperature of hot fluid, mass flow rate of cold fluid as inputs.[9] Considering the fact that fouling is a complex phenomena, each model developed to describe fouling either it be analytical model or machine learning model have their own strength and weaknesses. Considering this, a dynamic model based on Auto Regressive Integrated Moving Average with eXogenous inputs (ARIMA) is proposed in this paper for CF monitoring in reheater.

2. METHODOLOGY

In the present study an attempt is made to develop a dynamic regression model using ARIMAX to predict CF of reheater. The variation of CF with respect to time is very slow and hence ARIMA is chosen as it can well describe a slow drifting time series and hence applicable to slow changing inputs.[5] ARIMA model which is denoted by ARIMA (p, d, q) is a combination of AR and MA with a differencing, d to make the model stationary. p and q denotes the maximum time lag related to dependent variable and residual respectively. Exogenous variables added to ARIMA model results in ARIMAX model represented by ARIMAX (p, d, q) X(k). If the dependent variable is denoted by $\{Y_t\}$ and the exogenous variable set represented as $\{X_{1t}\}$, $\{X_{2t}\}$,..., $\{X_{kt}\}$, The linear expression to define ARIMAX model is,

$$\Delta^d Y_t = \delta + \sum_{i=1}^{k} \varphi_i X_{t-i} + \sum_{i=1}^{p} \mu_i Y_{t-i} + \sum_{i=1}^{q} \theta_i \varepsilon_{t-i} \quad \ldots (1)$$

Where φ_i, μ_i and θ_i denote the estimated coefficients, k represent the maximum time lag related to the input sequences and δ is a constant.

The steps involved in developing an ARIMA model are as follows:[4]

1. The time series of $\{X_t\}$ and $\{Y_t\}$ is examined for integrating type non-stationarity using the ACF plot and the unit-root tests. Unit root tests based on a null hypothesis of no unit root are preferred.
2. Determine the degree of differencing, d required to stationarize the time series by applying KPSS tests.
3. Determine the order of p and q in ARMA model by examining the sample ACF and PACF of the differenced series. Alternatively, p, q may be selected via an information criteria like AIC. The model with least value of the AIC is chosen to be the best model for forecasting.
4. After identifying a best fit ARIMA model, we have to check if the basic assumption of the model is satisfied i.e. if the model residual is zero mean Gaussian white noise process.

3. RESULTS AND DISCUSSION

Real time data obtained from M/s Neyveli Thermal power plant is used in this work. CF is not measured due to lack of instrumentation. Hence it is obtained by dual extended Kalman filter based state and parameter estimation[1] using the measured variables obtained from plant. The appropriate variables for CF monitoring as suggested by Anitha *et al.*[3] are flue gas input and output temperatures, steam input temperature, airflow, attemperator flow, coal flow.

3.1 ARIMAX Model Selection

In order to check if the gaussianity assumption in linear time series model holds, the histogram of CF series is examined which is shown in Figure 1a. It shows that gaussianity holds mildly with positive skewness of 0.3. This small deviation from ideality is ignored. The non-stationarity of the CF time series is analysed using the correlation plot which show a slow decay, suggesting the possibility of integrating effects coupled with stationary behaviour. In order to determine the number of differences required for eliminating the integrating effect, a unit root test is conducted which suggests the order of differencing to be 1. The differenced series and the corresponding correlation plot is shown in Figure 1. The ACF of the original and differenced series shown in Figure 1b and Figure 2b respectively are compared which shows that a first order differencing has eliminated the integrating effect. Similarly the optimal set of predictors is also differenced by an order 1 to make it stationary on mean.

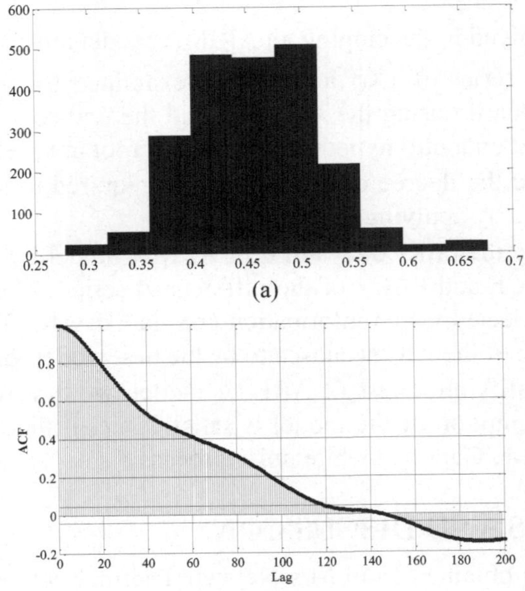

(a)

(b)

Fig. 1: a) Histogram of CF b) ACF of CF

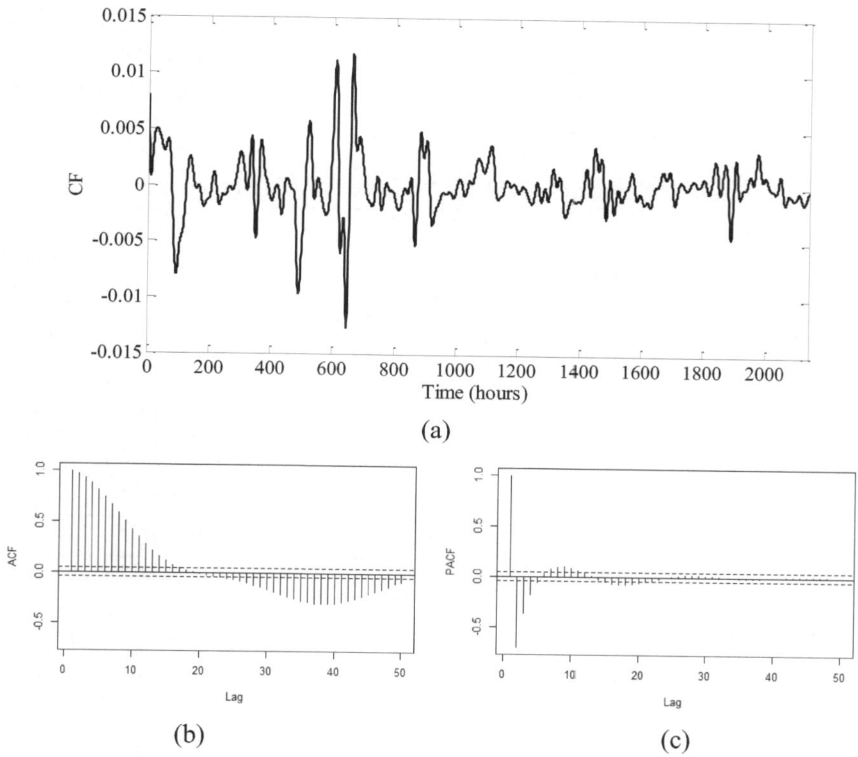

Fig. 2: (a) Differenced Series of CF (b) ACF (c) PACF

The data involving two soot blow cycles is used for model development and the remaining data set is used for testing. Table 2 presents the summary of different ARIMAX structures that is analysed. The number of lag periods necessary to make each predictor is determined by observing the cross-correlation between the corresponding stationarized predictor and the stationarized dependent variable. The cross-correlation of the predictors flue in, flue out, steam in, attemperator flow, coal flow with CF was found to be significant at lag 1, whereas the cross-correlation of air flow was found to be significant at lag 2. This suggests that flue in, flue out, steam in, attemperator flow, coal flow lagged by 1 period and air flow lagged by 2 periods might be a significant predictor of CF, the dependent variable. It was observed that all the selected model structures had uniform training error of 0.02. Whereas the testing RMSE is found to be minimum for ARIMAX (2, 1, 1) with Flue gas in, Flue gas out, Flue out with lag 1, air flow, coal flow and attemperator flow as predictor set. Hence considering the reasonable Akakie Information Criteria (AIC) in comparison with the other models and the low RMSE during testing that model structure ARIMAX (2, 1, 1) with 6 predictors is selected as the best fit. The estimated coefficients of the predictors, AR and MA terms of the best fit were also found significant using Z-test as tabulated in Table 1.

Table 1: Summary of ARIMAX Model Fitting Parameters

Model	*Fit*		*Predictors*	*Coefficient*	*Z-value*
	Testing RMSE	*AIC*			
ARIMAX (1, 1, 2)	0.78	–713.17	Flue in	2.65×10^{-3}	11.66
			Flue in (lag 2)	2.67×10^{-4}	1.28
			Flue out	-3.75×10^{-3}	-11.75
			Flue out (lag 2)	-7.78×10^{-4}	-2.63
			Steam in	1.34×10^{-4}	1.10
			Steam in (lag 2)	3.21×10^{-5}	0.24
			Air Flow	2.47×10^{-3}	11.24
			Coal Flow	2.63×10^{-3}	5.38
			Attemp Flow	3.39×10^{-3}	1.72
ARIMAX (2, 1, 1)	0.63	–719.21	Flue in	2.58×10^{-3}	11.71
			Flue out	-3.51×10^{-3}	–11.57
			Flue out (lag 2)	-3.89×10^{-4}	–3.05
			Air Flow	2.26×10^{-3}	11.46
			Coal Flow	2.58×10^{-3}	5.68
			Attemp Flow	4.57×10^{-3}	2.41
ARIMAX (2, 1, 1)	0.07	–704.46	Flue in	2.64×10^{-3}	11.83
			Flue out	3.40×10^{-3}	–10.91
			Flue out (lag 1)	1.70×10^{-3}	1.23
			Air Flow	2.17×10^{-3}	10.99
			Coal Flow	2.51×10^{-3}	5.47
			Attemp Flow	2.36×10^{-3}	2.79
ARIMAX (2, 1, 1)	0.39	–698.88	Flue in	2.81×10^{-3}	11.24
			Flue out	-3.87×10^{-3}	–11.80
			Air Flow	2.20×10^{-3}	9.83
			Coal Flow	2.88×10^{-3}	5.58
			Attemp Flow	5.05×10^{-3}	2.29

In time series modelling the residuals are assumed to be uncorrelated. The residual of the identified best fit ARIMAX (2, 1, 1) model is found to be uncorrelated which is confirmed by both ACF and Ljung-Box Q-test as shown in Figure 3. The p-values from lag 4 to lag 30 are shown which are found to be non-significant.

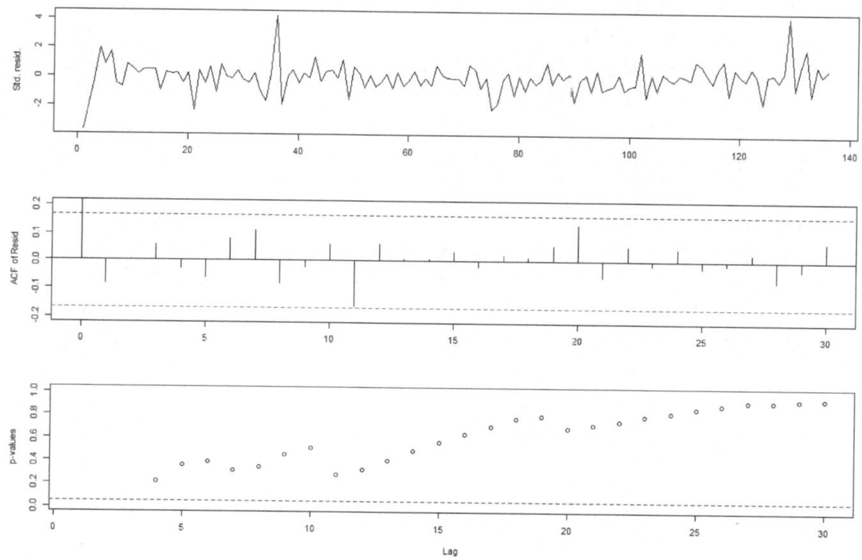

Fig. 3: Residual Diagnostics Plot

4. CONCLUSION

In this paper, ARIMAX model is developed with an aim to monitor CF with the optimal set of exogenous variables to the model. It is observed that ARIMAX model with a combination of two autoregressive terms, one order of differencing and one moving average term provides best fit. The residuals were also found to be uncorrelated which signifies that all the information related to the model has been retrieved.

ACKNOWLEDGMENT

This work is supported by UGC-SAP-DRS-I and DST-PURSE.

REFERENCES

[1] Anitha Kumari, S. and Srinivasan, S. (2018). "Extended Kalman filter for fouling detection in thermal power plant reheater", *Control Engineering Practice*, Vol. 73, pp. 91–99.

[2] AnithaKumari, S., Srinivasan, S. and Pugazhendi, R., Detection of ash fouling in thermal power plant, *National Academy science letters* 41 (2018), 369–373.

[3] Anitha Kumari, S. and Srinivasan, S., Ash fouling monitoring and soot blow optimization for reheater in thermal power plant, *Applied thermal engineering,* 149 (2019), 62–72.

[4] Wand, Y. and Xu, Y., “The application of ARIMAX model,” *Statistics and Decision* (2007), 9:1–1.

[5] Ashish, M. and Biao, H. (2002). Detection of abrupt change and applications in sensor decalibration monitoring. *ISA Transcations*, 155–166.

[6] Zhang, S., Shen, G., An, L., Niu, Y. and Jiang, G. (2015). “Monitoring ash fouling in power station boiler furnaces using acoustic pyrometry”, *Chemical Engineering Science*, Vol. 126, pp. 216–223.

[7] Delmotte, F., Dambrine, M., Delrot, S. and Lalot, S. (2013). “Fouling detection in a heat exchanger: A polynomial fuzzy observer approach”, *Control Engineering Practice*, Vol. 21, No. 10, pp. 1386–1395.

[8] Nasr, M.R. and Givi, M. (2006). “Modeling of crude oil fouling in preheat exchangers of refinery distillation units”, *Applied Thermal Engineering,* Vol. 26, pp. 1572–1577.

[9] Garcia, R.F. (2012). “Improving heat exchanger supervision using neural networks and rule based techniques”, *Expert Systems with Applications,* Vol. 39, pp. 3012–3021.

Development of Nonlinear State Space Model for ECG Signal Using Genetic Algorithm

S. Nambi Narayanan[1] and S. Sutha[2]
Department of Instrumentation engineering MIT Campus Anna University
E-mail: [1]vinoth2852@gmail.com; [2]sutha_mani@hotmail.com

ABSTRACT: *Analysis of ECG signal is very important for diagnosing cardiac vascular diseases (CVD) like cardiac arrhythmias, coronary heart diseases, cardiomyopathy. Modeling of ECG is very difficult due high nonlinearity and variability. This paper presents the modeling of nonlinear state space model for ECG signal using Genetic algorithm (GA). Three dimensional dynamic state space model is developed to generate synthetic ECG. Then the model parameters are tuned using GA to minimize the error between synthetic ECG and Real ECG signal. Finally the performances of the model are evaluated and compared with the conventional model. The comparison shows the feasibility of the model to generate various cardiac arrhythmias.*

Keywords: Synthetic ECG Signal, CVD, GA, Cardiac Arrhythmias.

1. INTRODUCTION

Dynamic model development of ECG is emerging research area in the biomedical field to analyse various heart diseases. Analysis of noisy real ECG signal is very difficult task due to improper selection of signal processing techniques. Hence synthetic ECG signal generation plays an important in physiological control system. Many researchers have developed mathematical models for generating synthetic ECG. Mcsharry *et al.* (2003) developed three dimensional dynamic state space model based on mean, standard deviation of heart rate, morphology of the PQRST cycle and power spectrum of RR tachogram. The model parameters are chosen to generate different morphologies for the PQRST-complex of ECG signal. Abdul Awal *et al.* (2011) proposed Gaussian wave base model which can simulate real time ECG wave with noise and compared it with real time ECG signal. He reported various noise signals involved in corrupting ECG signal. Das *et al.* developed fractional order model to generate ECG signal by modifying two identical Van-der Pol (VdP) family of oscillators. Mohammad Saad Billah et applied a novel method to model ECG beats using Gaussian fitting of order eight. They divided the ECG signal into two components as one above the base line and other one below the base line. The components are modeled separately and integrated together to get entire ECG signal. The model is compared with real ECG signal and performance are evaluated. Kianoush Nazarpour *et al.* (2007) introduced accurate mathematical model for fetal ECG based on sinc function and DC bias. The authors have taken model with 10 unknown parameters and tuned using genetic algorithm. Suranai Poungponsri *et al.* (2009) proposed ECG modeling based on wavelet neural network and is trained by a hybrid algorithm that includes the Adaptive Diversity Learning Particle Swarm Optimization (ADLPSO). Stuchilin *et al.* (2015) modeled ECG signal for various pathological conditions such as sinus arrhythmia, atrial, atrioventricular and ventricular extra systole, idioventricular rhythm, fibrillation and flutter of ventricles and atriums using Gaussian function. Sathya and Malathi (2013) developed Autoregressive technique for ECG modeling and coefficients of the model are trained using neural network.

The main aim of this paper is to develop nonlinear state space model for ECG signal using genetic algorithm. This paper is organized as follows: Section 2 describes the dynamic state space model. The GA based state space model is explained in Section 3 followed by performance analysis of the model in Section 4. Finally, the conclusions are presented in Section 5.

2. DYNAMIC STATE SPACE MODEL

Themotion equation of three dimensional dynamic state space model of synthetic ECG developed by Mcsharry *et al.* (2003) are presented in Equations (1)-(3):

$$\dot{x}(t) = \alpha x(t) - \omega y(t) \qquad \dots (1)$$

$$\dot{y}(t) = \alpha y(t) + \omega x(t) \qquad \dots (2)$$

$$\dot{z}(t) = -\sum_{i\in\{P,Q,R,S,T\}} a_i\, \Delta\theta_i\, exp\left(-\frac{\Delta\theta_i^2}{2b_i^2}\right)(z(t) - z_0(t)) \qquad \dots (3)$$

Where $\alpha = 1 - \sqrt{x^2 + y^2}$ $\Delta\theta_i = (\theta - \theta_i) \bmod 2\pi$, $\theta = a\tan2(y, x)$ (the four Quadrant arc tangent of the real parts of the element of x and y, with $-\pi \leq$ atan $2(y, x) \leq \pi$, and ω is the angular velocity of the trajectory as it moves around the limit cycle. Baseline wander was introduced by coupling the baseline value in z_0 in Equation (3) to the respiratory frequency f using $z_0(t) =$ Asin $(2\pi f\, t)$ where $A = 0.15\, mV$.

The normal ECG wave shown in Figure 1 is divided into P, Q, R, S, and T waves and described as follows:

P-wave: A small low-voltage deflection away from the baseline caused by the depolarization of the atria prior to atrial contraction as the activation (depolarization) wave-front propagates from the SA node through the atria.

PQ-interval: The time between the beginning of atrial depolarization and the beginning of ventricular depolarization.

QRS-complex: The largest-amplitude portion of the ECG, caused by currents generated when the ventricles depolarize prior to their contraction. Although atrial repolarization occurs before ventricular depolarization, the latter waveform (i.e. the QRS-complex) is of much greater amplitude and atrial repolarization is therefore not seen on the ECG.

QT-interval: The time between the onset of ventricular depolarization and the end of ventricular repolarization. Clinical studies have demonstrated that the QT-interval increases linearly as the RR-interval increases. Prolonged QT-interval may be associated with delayed ventricular repolarization which may cause ventricular tachyarrhythmias leading to sudden cardiac death.

ST-interval: The time between the end of S-wave and the beginning of T-wave. Significantly elevated or depressed amplitudes away from the baseline are often associated with cardiac illness.

T-wave: Ventricular repolarization, whereby the cardiac muscle is prepared for the next cycle of the ECG.

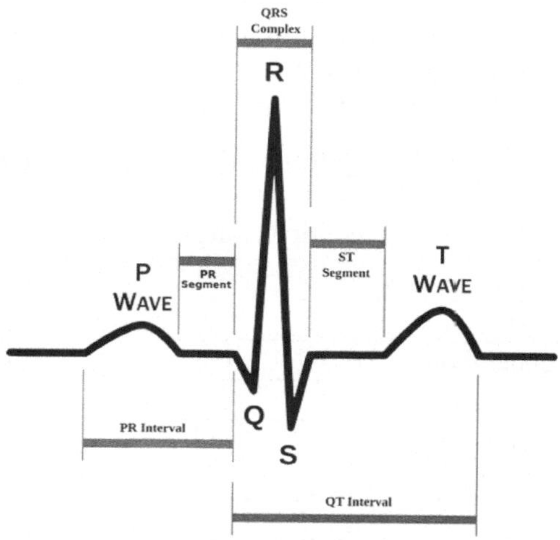

Fig. 1: Normal ECG Signal

The PQRST intervals are generated by three equations (1)–(3). The variations in the length of R-R interval is represented by angular velocity of ω(t).

Suppose $T(t)$ that represents the time series generated by the RR-process with power spectrum (f).

$$S(f) = \frac{\sigma_1^2}{\sqrt{2\pi c_1^2}} \exp\left(\frac{(f-f_1)^2}{2c_1^2}\right) + \frac{\sigma_2^2}{\sqrt{2\pi c_2^2}} \exp\left(\frac{(f-f_2)^2}{2c_2^2}\right) \quad \dots (4)$$

with mean f_1, f_2 and standard deviations c_1, c_2. Power in the LF and HF bands are given by σ_1^2 and σ_2^2respectively.whereas the variance equals the total area $\sigma^2 = \sigma_1^2 + \sigma_2^2$, yielding an LF/HF ratio of σ_1^2/σ_2^2. Its shows the power spectrum $S(f)$ given by $f_1 = 0.1$, $f_2 = 0.25$, $c_1 = 0.01$, $c_2 = 0.01$ and

$\frac{\sigma_1^2}{\sigma_2^2} = 0.5$. The Gaussian Frequency is motivated by the typical power spectrum of a real RR tach gram (4).

The time-dependent angular velocity of motion around the limit cycle is given by Equation (5).

$$\omega(t) = \frac{2\pi}{T(t)} \quad \dots (5)$$

3. ECG MODELING USING GA

Genetic algorithm (GA) is a heuristic search and optimisation technique inspired by natural evolution. They have been successfully applied to a wide range of real-world problems of significant complexity. The GA uses the following operators to manipulate the genetic composition of a population to achive the fitness function: reproduction, crossover and mutation. The main objective function of the modeling to minimize the least square error between Real ECG model taken form,

$$\text{MITBIH Fitness Function} = \sum(\text{Real ECG} - \text{Synthetic ECG})^2 \quad \dots (6)$$

Table 1: GA Parameters used for Tuning GA based IMC-PI Scheme

Parameters	*Value*
Population size	15
Number of generation	20
Probability of crossover	0.99
Probability of mutation	0.01
Selection procedure	Ranking
Crossover type	Single point crossover
Objective function	ISE

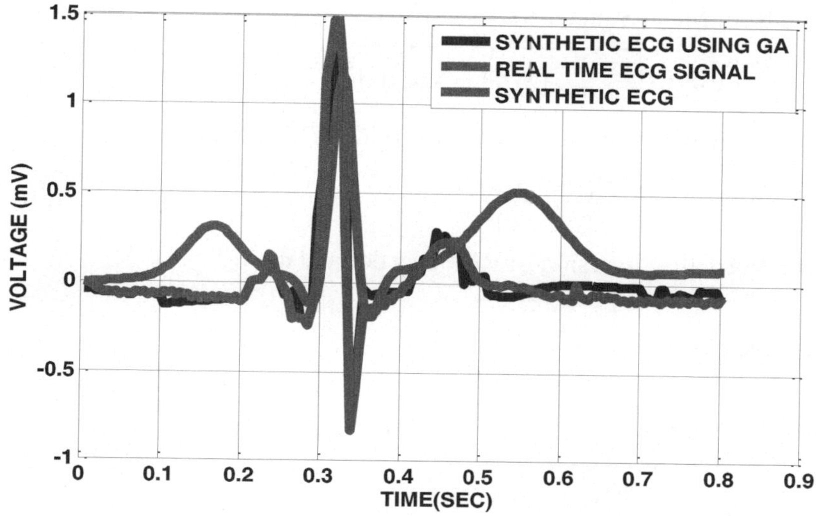

Fig. 2: Comparison of Synthetic ECG Signal with and Real ECG Signal

From the Figure 2, it is found that the response of synthetic ECG signal generated by model parameters tuned using GA is very close to real time signal than existing model response without GA.

Table 2: Model Parameters of Synthetic ECG

ECG Model	*Model Parameters*	*P*	*Q*	*R*	*S*	*T*
Synthetic ECG	t_i	–70	–15	0	15	100
	a_i	1.2	–5	30	–7.5	0.75
	b_i	0.25	0.1	0.1	0.1	0.4
Synthetic ECG with GA	t_i	–6.83	–4.18	0	14.4	96.49
	a_i	0.189	–4.85	28.70	–3.64	0.6
	b_i	0.204	0.091	0.13	0.091	0.253

4. PERFORMANCE ANALYSIS OF THE MODEL

The goodness of the model fitting is measured by the following performance measures:

The Mean Square Error (MSE) is defined as,

$$\text{MSE} = \frac{1}{N}\sum_{n=0}^{N-1}[x(n) - x_m(n)]^2 \qquad \dots (7)$$

The Root Mean Square Error (RMSE) is defined as,

$$\text{RMSE} = \sqrt{\frac{1}{N}\sum_{n=0}^{N-1}[x(n) - x_m(n)]^2} \qquad \dots (8)$$

The Normalised Mean Square error (NMSE) is defined as,

$$\text{NMSE} = \frac{\sum_{n=0}^{N-1}[x(n)-x_m(n)]^2}{\sum_{n=0}^{N-1}[x(n)]^2} \qquad \dots (9)$$

The Normalised version of RMSE is defined as,

$$\text{NRMSE} = \sqrt{\frac{\sum_{n=0}^{N-1}[x(n)-x_m(n)]^2}{\sum_{n=0}^{N-1}[x(n)]^2}} \qquad \dots (10)$$

Percent Root Mean square Difference (PRD) is defined as,

$$\text{PRD} = \sqrt{\frac{\sum_{n=0}^{N-1}[x(n)-x_m(n)]^2}{\sum_{n=0}^{N-1}[x(n)]^2}} \times 100\% \qquad \dots (11)$$

The performance parameters of the exiting model and proposed model are tabulated in Table 3. From the table it is observed that synthetic ECG using GA provides very smaller values of MSE, NMSE, RMSE, NRMSE and PRD than existing model.

Table 3: Performance Measures of ECG Model

Goodness of Fitting	*Synthetic ECG*	*Synthetic ECG using GA*
MSE	0.1318	0.0085
NMSE	1.5982	0.1640
RMSE	0.3630	0.0924
NRMSE	1.2642	0.4049
PRD	126.42	40.49

5. CONCLUSION

Dynamic state space model is developed for generating synthetic ECG signal for normal person. The model parameters are tuned to achieve real ECG using least square algorithm. Then the performance of the proposed model is evaluated to determine the Goodness of the fitting. From the Table 3, it is concluded that dynamic model using GA gives Good fitness values than conventional model. The work is extended to generate synthetic ECG for various pathological conditions.

REFERENCES

[1] McSharry, Patrick E.; Clifford, Gari D.; Tarassenko, Lionel and Leonard, A. "A Dynamical Model for Generating Synthetic Electrocardiogram Signals", *IEEE Transaction on Biomedical Engineering*, Vol. 50, No. 3, March 2003.

[2] Awall, Md. Abdul; Mostafa, Sheikh Shanawaz and Ahmad, Mohiuddin, "Simplified Mathematical Model for Generating ECD Signal and Fitting the Model Using Nonlinear Least Square Technique", *Proceedings of the International Conference on Mechanical Engineering 2011 (ICME2011)*, 18–20 December, 2011.

[3] Das, Saptarshi and Maharatna, Koushik, "Fractional Dynamical Model for the Generation of ECG like Signals from Filtered Coupled Van-der Pol Oscillators", *Computer methods and programs in biomedicine*, pp. 112, August 2013.

[4] Billah, Mohammad Saad; Mahmud, Tahmida Binte; Snigdha, Farhana Sharmin and Arafat, Muhammad Abdullah, "A Novel Method to Model ECG Beats using Gaussian Functions", *4th International Conference on Biomedical Engineering and Informatics (BMEI)*, 2011.

[5] Nazarpour, Kianoush; Ebadi, Siamak and Sanei, Saeid, "Fetal Electrocardiogram Signal Modelling Using Genetic Algorithm", *International Workshop on Medical Measurements and Applications*, 2007.

[6] Poungponsri, Suranai and Yu, Xiao-Hua, "Electrocardiogram (ECG) Signal Modeling and Noise Reduction Using Wavelet Neural Networks", *Proceedings of the IEEE International Conference on Automation and Logistics*, 2009.

[7] Stuchilin, V.V., Rumyantsevaand, V.A. and Svirin, I.S., "Developing Electrocardiogram Mathematical Model for Cardiovascular Pathological Conditions and Cardiac Arrhythmia", *Indian Journal of Science and Technology*, Vol. 8, 2015.

[8] Sathya, N. and Malathi, R., "Modeling of ECG Signal and Validation by Neural Networks", *Recent Advancements in System Modelling Applications*, 2013.

[9] Kaur, Amanjyot and Suma, Anita, "Disease Classification Using ECG Signal Based on PCA Feature Along with GA & ANN Classifier", *International Research Journal of Engineering and Technology (IRJET)*, 2017.

[10] McSharry, Patrick E.; Clifford, Gari D.; Tarassenko, Lionel and Leonard, A., "Advanced Methods and Tools for ECG Data Analysis", *Engineering in Medicine and Biology*, 2006.

[11] Omran, Safaa S.; Taha, S.M.R. and Awadh, Nassr Ali, "ECG Rhythm Analysis by Using Neuro-Genetic Algorithms", *Journal of Basic and Applied Sciences*, Vol. 1, pp. 3, October 2009.

Study and Analysis of Multimodal Biometrics

T. Sathish Kumar[1] and K. Trinayani[2]

[1]SES, SISD, Electronics and Instrumentation Group, IGCAR, Kalpakkam

[2]M.E. Instrumentation Engineering, MIT Campus, Anna University, Chennai

[2]E-mail: trinayanikmisra@gmail.com

ABSTRACT: *Biometric based authentication systems are used in a wide range of applications due to the permanent and non-repudiation nature of the biometrics. However, high failure to enroll ratio and absence of a particular biometric among some members in the population pose challenges for adoption of biometric systems. Further, confidence among users about the accuracy of a biometric system determines the adoption of biometric systems. To address these challenges, multimodal biometrics was proposed. This work is a study of score level fusion of face, left and right finger multimodal biometrics. Experiments were conducted on NIST multimodal score datasets. Through experiments, we demonstrate that multimodal biometrics gives lower EER (Equal Error Rate) than Unimodal Biometrics. We proceed to determine the normalization and fusion techniques that yields the lowest EER.*

Keywords: Multimodal Biometrics, Support Vector Machines, Score-level fusion, False Acceptance Rate, False Rejection Rate, Genuine Acceptance Rate, Equal Error.

1. INTRODUCTION

Biometrics is the science of establishing the identity of an individual based on the physical, chemical or behavioral attributes of the person.[1] Biometrics are classified into physiological and behavioral biometrics. The physiological characteristics include iris, face, fingerprint, hand geometry, palm print etc. The behavioral characteristics include gait, voice, and typing rhythm etc. Biometric systems record unique features from a given biometric image and use those for authentication.

Biometric systems which rely on only one type of biometric are called as unimodal biometric systems. A unimodal system has several disadvantages[1] which make it undesirable for large scale application. These disadvantages are listed below.

- Noise in the sensed biometric input (due to the sensor or due to the nature of biometric).

- Non-universal nature of the biometric. E.g.: Hand geometry based biometric systems could not enroll differently abled personnel.
- Limited Accuracy.
- Prone to security attacks.

A multimodal system uses two or more biometric features of a person for authentication.[1] Different biometric inputs are fused together for deciding the authenticity of the personnel.

Biometric fusion is the process combining multiple biometric inputs into a single fused vector for authentication. The fusion can happen in four different levels.[3]

- *Sensor Level fusion:* involves fusing the biometric inputs at the sensor.
- *Feature Level fusion:* involves fusing feature vectors from various biometric units into one feature vector.[3]
- *Matching Score Level fusion:* involves fusing the similarity or dissimilarity scores of various biometric units.[4,5]
- *Decision Level fusion:* involves making a single decision among a few options.

The objective of the work is to study various normalization and fusion techniques for score based fusion. Based on the study, we arrive at the best combination of normalization and fusion techniques that yield the lowest EER.

2. BIOMETRIC SYSTEM BASICS

The following terminologies are used for measuring the efficiency of biometric based authentication systems.

- *FAR* : False Acceptance Ratio.
- *FRR* : False Rejection Ratio.
- *GAR* : Genuine Acceptance Ratio
- *GRR* : Genuine Rejection Ratio.
- *EER* : Equal Error Rate.
- *FAR* : It is the number of Impostors that the biometric system accepts as a Genuine User.
- *FRR* : It is the number of genuine users that the biometric system rejects as an Impostor.
- *GAR* : It is the number of genuine users that the biometric system accepts as genuine.
- *GRR* : It is the number of Impostors that the biometric system rejects as Impostors.
- *EER* : It is the rate of error at which the FAR equals FRR.

In the case of unimodal biometrics, FAR and the FRR tends to be higher leading to high EER.

3. LITERATURE REVIEW

The basics of biometric systems with their advantages, disadvantages and the types of biometric systems are detailed in.[1] Basic biometric features used in today's biometric systems and the need for multimodal biometric systems, along with the various normalization schemes and a case study of Sum-Rule and SVM based fusion using two kernels is given in.[2] The paper[3] deals with various algorithms like, extraction fingerprint biometric using histogram of gradients, palm print extraction using combined PCA and LDA and hand geometry feature extraction using Harris Corner Detection Algorithm and feature level fusion by concatenation and SVM based fusion. Another variation of using multimodal biometrics is performed by the fusion of fingerprint and iris recognition and also, methods of minutiae matching and feature extraction for fingerprint biometric is given in.[4] Fusion using Likelihood ratios and SVM, LRR and GVM models have been implemented in.[5] One of the types of fusion focussed in this researchpaper is the matching-score level fusion by using Sum-rule and its variants as detailed in.[6] One variant of the rule-based schemes is User-Weighted fusion scheme. According to literature,[6] it has been found to give optimal results when weights are in multiples of 0.1 and hence not emphasized in this research work. Various machine learning algorithms can also be applied for the fusion. The simplest including[7] Decision trees and LDA haved paved the way for SVM based fusion.[2,5] From[2,7] we get an idea of how multimodal biometrics can involve certain machine learning methods. The real-world appplications of machine learning in basic classification and regression problems are deatiled in.[8] The fusions discussed in this research paper are Sum-rule based and SVM-based with four kernels.

4. PROPOSED WORK

We study the various score level fusion techniques by applying them on National Institute of Standards and Technology (NIST), USA multimodal database, NIST Biometrics Score Set - BSSR1 [Web link: https://www.nist.gov/itl/iad/image-group/nist-biometric-scores-set-bssr1]. The database contains scores for genuine and imposter matching for 517 users for three different biometrics namely face, left fingerprint and right fingerprints.

We first calculate EER for unimodal biometrics using the genuine and imposter scores. Then we proceed to apply two separate fusion methodologies to these scores. Rule based fusion is described in section V and

SVM based fusion is described in section VI. We draw distribution graphs and calculate EER in each case to determine accuracy.

5. RULE-BASED FUSION

The sum rule based methods, require a preprocessing step. This step is called as normalization. After normalization, fusion follows.

5.1 Normalization

This is the first step in Rule-based fusion methods. Normalization is done to map the different ranges of scores obtained from different biometric units into a single interval. This is done for the ease of fusion. Let X denote the set of raw matching scores from a specific matcher and let x ε X'. The normalized score of X is x'. The types of Normalization Schemes that are discussed include, Min-Max, Z-score, RHE (Reduction of High scores Effect).

1. *Min-Max Normalization:* This normalization maps the raw matching scores to the interval (0, 1) and retains the original distribution of matching scores.[2]

$$x' = (x - \min(X))/(\max(X) - \min(X)) \quad \dots (1)$$

Where, min(X) - minimum value of raw matching score and, max(X) - maximum value of raw matching score.

2. *Z-Score Normalization:* Calculates normalized scores using arithmetic mean and standard deviation of given data.[2, 5]

$$x' = (x - \text{mean}(X))/\text{Std}(X) \quad \dots (2)$$

where, Mean(X) - arithmetic mean of X and Std (X) - standard deviation of X.

3. *RHE-Normalization:* This scheme was implemented from paper.[1] It is a variant of Min-Max Normalization that is considered more efficient.[2, 6]

$$x' = (x - \min(X))/((\text{mean}(X^*) + \text{std}(X^*)) - \min(X)) \quad \dots (3)$$

where, x = Raw Genuine/Impostor score, X = Total score set, X* = Distribution of genuine scores, x' = Normalized score.

Extremely high scores or outliers can affect the distribution of data in large databases.[2] Hence, we add the mean of genuine scores with the standard deviation instead of the maximum value of all scores to avoid this.

We take the scores of 517 users from the NIST, USA database. There are 517 genuine and 517 impostor scores for each users' one biometric trait. The traits considered for the analysis are face, left fingerprint and right fingerprint scores. There initial distribution is plotted for genuine and impostor scores as, their range, VS the count of the scores within the range as given in Figures 1–6.

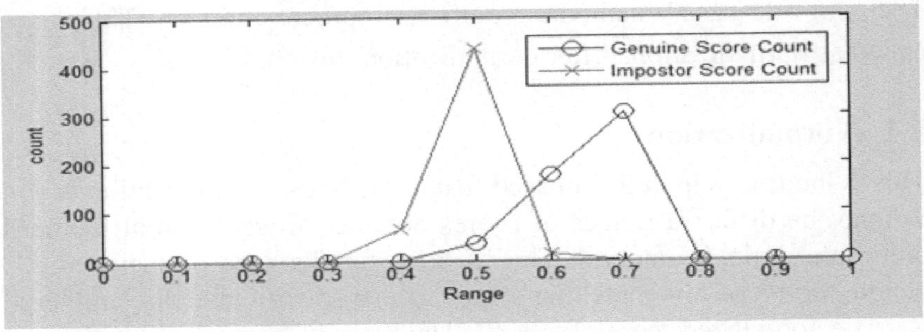

Fig. 1: Pre-Normalized Distribution of Scores (Face)

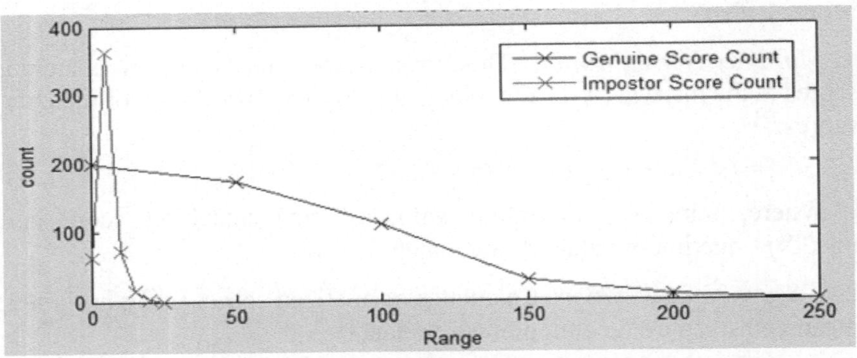

Fig. 2: Pre-Normalized Distribution of Scores (Left Fingerprint)

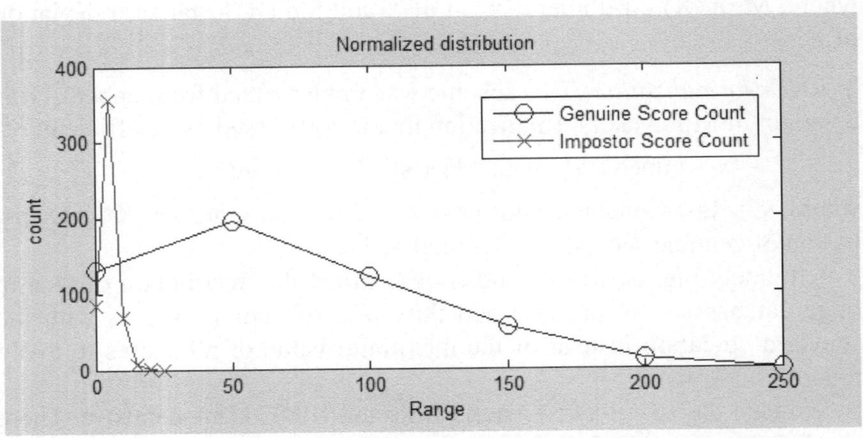

Fig. 3: Pre-Normalized Distribution of Scores (Right Fingerprint)

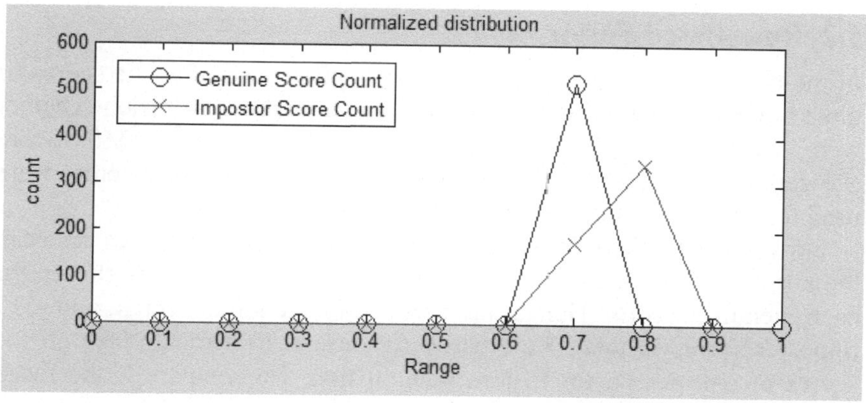

Fig. 4: Min-Max Normalized Distribution of Scores (Face)

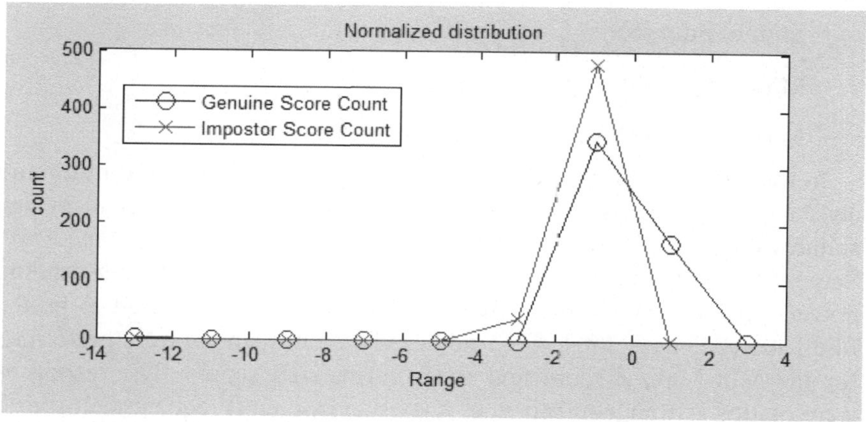

Fig. 5: Z-Score Normalized Distribution of Scores (Left Fingerprint)

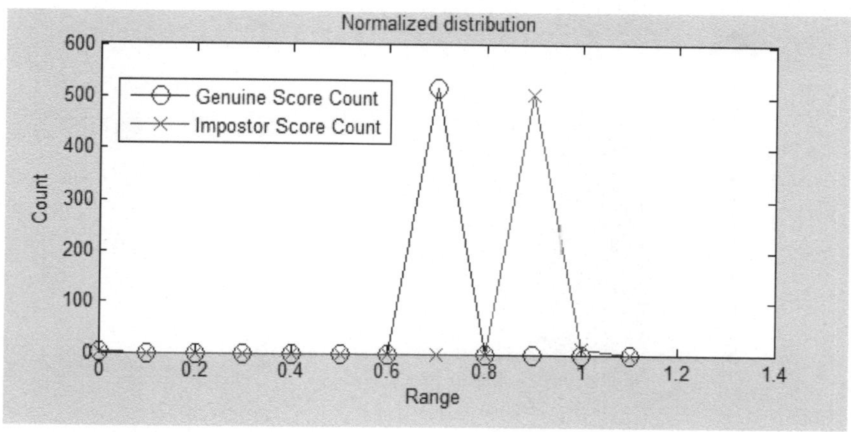

Fig. 6: RHE Normalized Distribution of Scores (Right Fingerprint)

5.2 Rule-Based Fusion

In the case of Rule-based fusion, fusion is performed after the scores of three biometrics are normalized to the same interval. Five fusion schemes were used namely, Sum-Rule based,[2] MIN-Score,[2,6] MIN-Score (Powers),[2,6] MAS-Score,[6] MAS-Score (Powers).[6] These schemes were used for 517 users of the NIST, USA database.

For each user, there is one genuine and 516 imposter scores for each biometric. For the purpose of this study, one imposter score is chosen for each genuine score. Hence there are 517 genuine scores and 517 imposter scores in total. Rule based fusion and its variants take all the scores of one person for fusion. After fusion, we compare if the fused score is above or below a certain threshold and then authenticate or reject.

- Simple-Sum (SS): $f_i = \sum_{m=1}^{M} n_i^m, V_i$
- Min-Score (MIS): $f_i = \min(n_i^1, n_i^2, ..., n_i^M), V_i$
- Max-Score (MAS): $f_i = \max(n_i^1, n_i^2, ..., n_i^M), V_i$

In the case of Min-Score (MIS) and Max-Score (MAS) the minimum and maximum scores among the biometrics for one user is taken and suitably considered as the fused score. For Min-Score (Powers) and Max-Score (Powers), the individual biometrics are considered in ascending powers and the minimum or maximum of those is considered to be the fused score. Thus in total, five rule-based fusion methods were performed for the Min-Max, Z-score and RHE normalized scores. The responses were plotted as Range of scores in X-axis and Count (Frequency) of scores occurring within the particular range as in Figure 7.

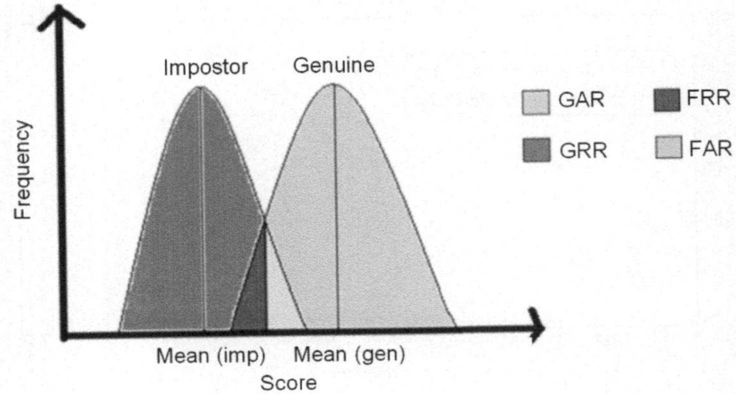

Fig. 7: Evaluation Parameters

5.3 Per -Evaluation of Rule-Based Fusion

FAR Genuine/Total no. of Impostor scores (1)
F Impostor/Total no. of Genuine scores (2)
cepted/Total no. of Genuine scores (3)
ejected/Total no. of Impostor scores (4).

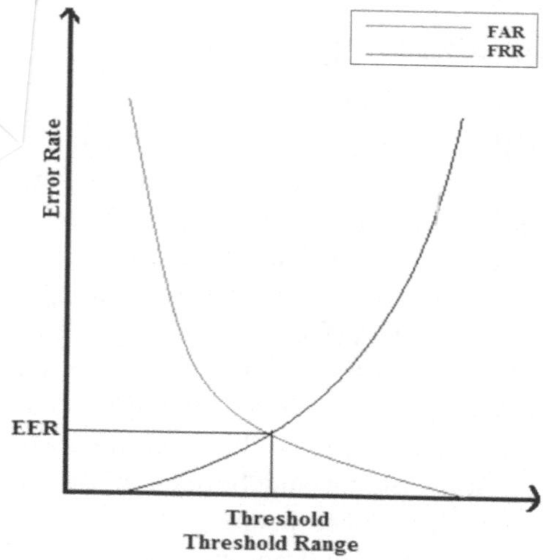

Fig. 8: Determination of EER and Threshold

The EER and threshold are found by plotting FAR and FRR (Error Rate) vs the Threshold as shown in Figure 8.

6. SVM-BASED FUSION

SVM is the abbreviation for Support Vector Machine. It is a supervised machine learning algorithm that uses Support Vectors (SV) to have a decision boundary that separates two sets of data via a hyper plane.[8] Machine learning can be defined as follows, it is a computer program which learns from experience E with respect to some class of tasks T and performance measure P, if its performance measure in tasks T, as measured by P, improves with experience E.[8]

6.1 SVM for Fusion

The SVM is a supervised machine learning algorithm that trains to learn things in the given test data using labels and builds a predictive model. It then tries to predict test data into the correct labels. Support Vectors are parts of the data which can be used to define a boundary in order to classify the data. This was done via Matlab 2015a so... specific SVM functions such as 'fitcsvm', 'svmtrain', 'sv... from its function library. The SVM makes use of kern... Kernel functions are those that transform non-linear data to ... The kernel functions which were used are Linear kernel, Ra... function or Gaussian kernel, Quadratic kernel and Polynomial ke...

6.2 Performance of SVM-Based Fusion

In this research work SVM with these four kernels were applied on 517 users of the NIST, USA database. The individual scores of each ... the users' three biometrics were taken and trained. The scores for which the SVM was applied were Raw scores, Min-Max normalized scores, RHE Normalized scores and Z-normalized Scores.

Of the 517 users, 100 users were used for training the SVM and 417 users were used as testing data (face, left and right fingerprint scores). In order to test the efficiency of the training algorithm, the testing data was selected at random and used for testing. The testing data used was 20% of the total 517 users, i.e., 104 users. This was done as random 104 user data selected at the top, middle and bottom of the database. This was done for raw scores, min-max scores, rhe scores and z-scores by applying linear, rbf, quadratic and polynomial kernels for each of them. The support vector coordinates, obtained so were plotted in 3-D along with the genuine and the impostor scores. They were evaluated in terms of test accuracy, correct real prediction and correct false prediction. The results are in terms of Test accuracy, FAR, FRR, GAR, GRR. The readings have been tabulated. In Tables 5–7.

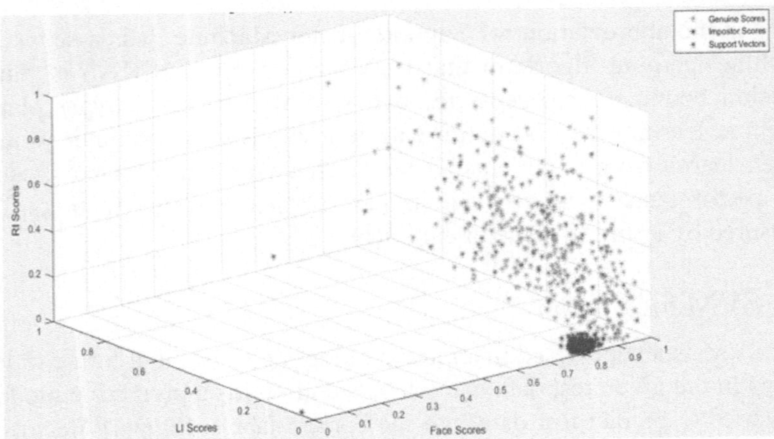

Fig. 9: Testing Data & SV Distribution for Linear Kernel Min-Max Scores

7. RESULTS

7.1 EER's of Unimodal and Multimodal Biometric System

EER were calculated for both the matching scores of unimodal and multimodal rule-based fused scores. For Unimodal system, the raw scores of face, left fingerprint and right fingerprint were plotted as in Figure 1 and Performance evaluation was done as per Figure 2. For Multimodal Biometric systems, performance evaluation was done on the three types of normalized scores in each rule-based fusion method.

Table 1: EER of a Unimodal Biometric System

S. No.	*Type of Raw Score*	*EER*
1.	Face	0.0432
2.	Right-Index Fingerprint	0.0434
3.	Left-Index Fingerprint	0.0824

Table 2: EER of Min-Max Normalized and Fused Scores

S. No.	*Method (Fusion)*	*EER*	*Threshold*
1.	Sum-Rule	0.001343	0.923
2.	Max-Score	0.045	0.836
3.	Max-Score(Powers)	0.0425	0.836

Table 3: EER of Z-Score Normalized and Fused Scores

S. No.	*Method (Fusion)*	*EER*	*Threshold*
1.	Sum-Rule	0.0174	0.342
2.	Max-Score	0.01	0.075
3.	Max-Score(Powers)	0.1315	0.545
4.	Min-Score	0.0614	0.6983
5.	Min-Score(Powers)	0.0437	0.4007

Table 4: EER of RHE Normalized and Fused Scores

S. No.	*Method (Fusion)*	*EER*	*Threshold*
1.	Sum-Rule	0.0135	1.044
2.	Max-Score	0.0356	0.8738
3.	Max-Score (Powers)	0.0373	0.8743
4.	Min-Score	0.0746	0.0159

It is to be noted that for min-max normalized scores, Min-score and Min-Score (Powers) based fusion is not performed. This is due to the fact that the minimum genuine and impostor scores of the three matchers for all the 517 individuals are convergent in a small interval and hence fixing thresholds is difficult. The RHE normalized scores too followed a similar pattern and hence Min-Score (Powers) method was not applied.

7.2 ROC (Receiver Operating Characteristics) Curves for Performance Comparison of Rule-Based Fusion

ROC curves are plotted as true positive vs false positive for the purpose of evaluating or comparing the performances of various methods used in a field. The various rule-based fusion methods were plotted as GAR vs FAR. The results thus obtained are given in Figures 10–12.

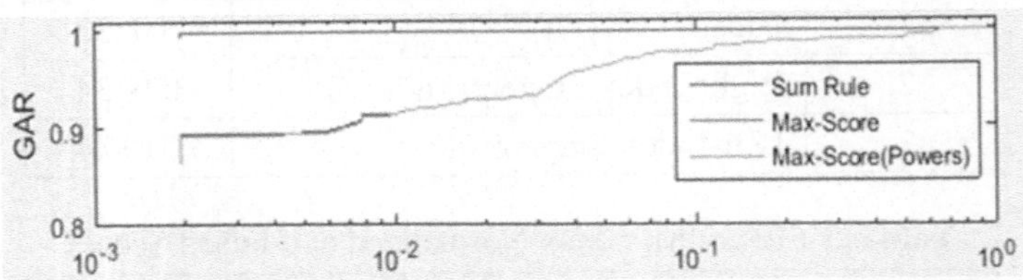

Fig. 10: ROC Curves of Min-Max Normalized and Fused Scores

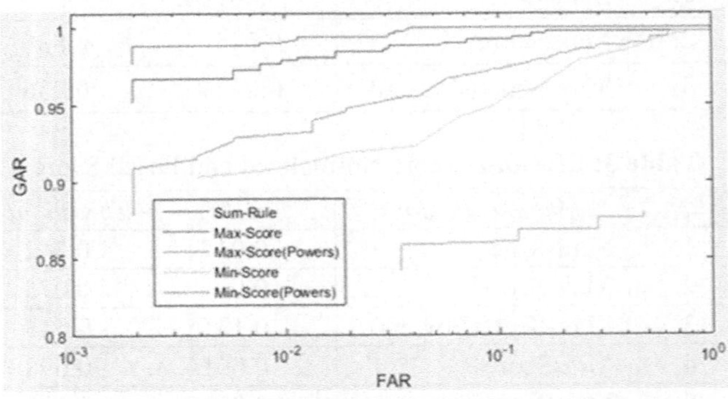

Fig. 11: ROC Curves of Z—Normalized and Fused Scores

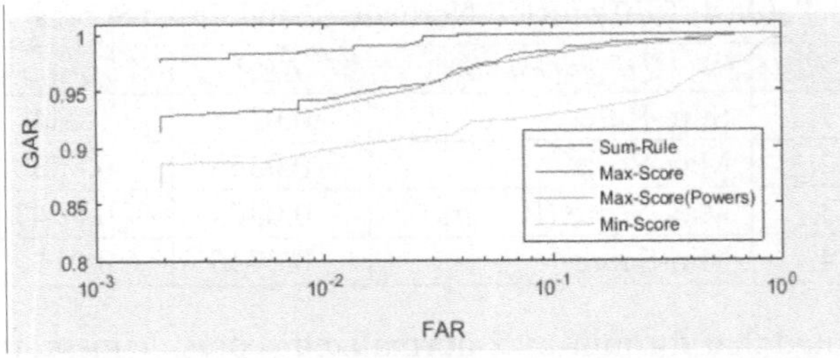

Fig. 12: ROC Curves of RHE—Normalized and Fused Scores

7.3 Results of SVM-Based Fusion

Of the 517 users, 100 users were used for training the SVM and 417 users were used as testing data. In order to test the efficiency of the training algorithm, the testing data was selected at random and used for testing. The testing data used was 20% of the total 517 users, i.e., 104 users. This was done as random 104 user data selected at the top, middle and bottom of the database for raw scores, min-max scores, rhe scores and Z-scores by applying linear, rbf, quadratic and polynomial kernels. They were evaluated in terms of accuracy, correct real prediction and correct false prediction. The results are in terms of test accuracy, FAR, FRR, GAR, GRR.

Table 5: Top 20% of Scores

Kernel	*Score Type*	*Accuracy*	*GAR*	*GRR*	*FAR*	*FRR*
Linear	Raw	99.04	0.9807	1	0	0.0193
	Min-Max	99.04	0.9807	1	0	0.0193
	Z	99.04	0.9807	1	0	0.0193
	RHE	90.38	1	0.8076	0.1924	0
RBF	Raw	99.04	0.9807	1	0	0.0193
	Min-Max	99.04	0.9807	1	0	0.0193
	Z	99.04	0.9807	1	0	0.0193
	RHE	90.38	1	0.8076	0.1924	0
Quadratic	Raw	100	1	1	0	0.0193
	Min-Max	100	1	1	0	0.0193
	Z	99.04	0.9807	1	0	0.0193
	RHE	90.38	1	0.8076	0.1924	0
Polynomial	Raw	100	1	1	0	0
	Min-Max	100	1	1	0	0
	Z	100	1	1	0	0
	RHE	100	1	0.8076	0.1924	0

Table 6: Middle 20% of Scores

Kernel	*Score Type*	*Accuracy*	*GAR*	*GRR*	*FAR*	*FRR*
Linear	Raw	98.56	0.9711	1	0	0.0289
	Min-Max	99.52	0.9903	1	0	0.0097
	Z	99.04	0.9807	1	0	0.0193
	RHE	100	1	1	0	0
RBF	Raw	98.08	0.9615	1	0	0.0385
	Min-Max	100	1	1	0	0
	Z	98.56	0.9711	1	0	0.0289
	RHE	100	1	1	0	0
Quadratic	Raw	98.56	0.9711	1	0	0.0289
	Min-Max	100	1	1	0	0
	Z	98.56	0.9711	1	0	0.0289
	RHE	100	1	1	0	0
Polynomial	Raw	98.56	0.9711	1	0	0.0289
	Min-Max	100	1	1	0	0
	Z	99.04	0.9807	1	0	0.0193
	RHE	100	1	1	0	0

Table 7: Bottom 20% of Scores

Kernel	*Score Type*	*Accuracy*	*GAR*	*GRR*	*FAR*	*FRR*
Linear	Raw	100	1	1	0	0
	Min-Max	100	1	1	0	0
	Z	100	1	1	0	0
	RHE	100	1	1	0	0
RBF	Raw	100	1	1	0	0
	Min-Max	100	1	1	0	0
	Z	100	1	1	0	0
	RHE	100	1	1	0	0
Quadratic	Raw	100	1	1	0	0
	Min-Max	100	1	1	0	0
	Z	100	1	1	0	0
	RHE	100	1	1	0	0
Polynomial	Raw	100	1	1	0	0
	Min-Max	100	1	1	0	0
	Z	100	1	1	0	0
	RHE	100	1	1	0	0

8. CONCLUSION

Based on the performance evaluation performed for the raw scores of a unimodal biometric system and it is found that The EER varies from 4.32% to 8.24%. From ROC curves and performance evaluation of various rule-based fusion methods, we are able to determine that among Min-Max normalized and fused scores that, Simple Sum-Rule based fusion yields the lowest EER of 0.13%. Among Z-Score normalized and fused scores, 0.1% EER is obtained by Max-Scores method. Among RHE normalized and fused scores, Sum-rule based fusion has an EER of 1.35%. On comparing unimodal and multimodal biometric systems we are able to conclude that while unimodal has an EER minimum of 4.32% and maximum 8.24%, a multimodal biometric system with Sum-rule based fusion preceded by Min-Max normalization score yields lowest EER of 0.13%. It can decisively be said that a multimodal biometric system is better performing than a unimodal biometric system. For SVM based fusion, polynomial kernel gives the best accuracy of 98.56% to 100% with great consistency over the complete database.

ACKNOWLEDGEMENT

We thank Indira Gandhi Center for Atomic Research, Kalpakkam for providing the resources and infrastructure to carry out this research work.

REFERENCES

[1] Anil, K. Jain; Patrick, Flynn and Arun, A. Ross, Handbook of Biometrics, Springer, 2008.

[2] Minxing, He; Shi-Jinn, Horng; Pingzhi, Fan; Ray-Shine, Run; Rong-Jian, Jui-Lin Lai, Muhammad Khurram Khan, Kevin Octavious Sentosa, Performance Evaluation of Score Level Fusion in Multimodal Biometric Systems, *Pattern Recognition,* 43 (2010), Elsevier, 1789–1800.

[3] Vinodkumar, Srikantaswamy R., Fusion of Fingerprint, Palmprint and Hand Geometry for an Efficient Multimodal person authentication system, *IEEE, 2nd International Conference on Applied and Theoretical Computing and Communication Technology (iCATccT),* 2016.

[4] Chandeesh Babu, K.R.; Parkavi, R. and Ajeeth Kumar, J., Multimodal Biometrics for User Authentication, *IEEE, 11th International Conference on Intelligent Systems and Control (ISCO),* 2017.

[5] Qian, Tao and Raymond, Veldhuis, Robust Biometric Score Fusion by Naïve Likelihood Ratio via Receiver Operating Characteristics, *IEEE Transactions on Information Forensics and Security,* Vol. 8, No. 2, February 2013.

[6] Subash, V. Thul; Anurag, Rishishwar and Neetesh, Raghuwansh, Sum Rule Based Matching Score Level Fusion of Fingerprint and Iris Images for Multimodal Biometrics Identification, *International Research Journal of Engineering and Technology*, Vol. 03, Issue 02, Feb. 2016.

[7] Arun, Ross and Anil, Jain, Information Fusion in Biometrics, *Patten Recognition Letters*, Vol. 24, Issue 13, Feb. 2003.

[8] Tom, Mitchell, Machine Learning, McGrawHill, Issue 1, March, 1997.

Plant Species Recognition Using Modified LeNet-5 CNN Architecture

S. Anubha Pearline[1] and V. Sathiesh Kumar[2]
Department of Electronics Engineering, MIT Campus,
Anna University, Chennai, India
E-mail: [1]anubhapearl@gmail.com; [2]sathieshkumar@annauniv.edu

ABSTRACT: *Plants play a vital role in preserving the ecosystem. It is used as food, medicine, furniture material, etc. Image processing methods stand as a prominent visual method for identification and recognition of plant species. Traditional image processing methods require several steps to carry out recognition tasks (Feature Extraction and Classification). In recent years, image processing and machine learning are overpowered by deep learning methods. Particularly among deep learning methods, image processing is carried out using Convolutional Neural Network (CNN). In this paper, plant species recognition using a modified LeNet-5 deep neural network is proposed. It uses convolution, max pooling and fully connected layers. Convolution layers are used with Rectified Linear Unit (ReLU) activation function. Adam optimizer is used for training the model. Our deep network is analyzed with the help of benchmark dataset, Flavia containing 32 leaf classes and obtained top-1 and top-5 accuracy of 89.58% and 98.96%, respectively.*

Keywords: Plant Species Recognition, Plant Classification, Convolutional Neural Networks, Deep Learning.

1. INTRODUCTION

A large variety of plants is present all over the world. Among them, there exist consumable and non-consumable plants. Several consumable plants are medicinally useful. These consumable plant acts as a primary producer in the food chain. Thus, plants are a rich source of human beings and animals. Over the years, plant production is shrunken due to the encroachment of buildings such as companies, houses, apartments etc. Many young people do not know about the consumable plants that are medicinally useful available in our surroundings. Therefore, it is highly important to recognize and identify such plant species. Commonly, recognizing plants are carried out through image processing methods. The literature on plant recognition is surveyed and is discussed below.

Zhang *et al.*[1] proposed a method for plant recognition by fusing Singular Value Decomposition (SVD) and sparse representation together. The authors report that plant classification is important in protecting ecological environment and plant species Instead of extracting features from images, based on sparse coefficients the dimensionality of the images are reduced. Without training an over-complete dictionary is formed from the plant images. Class-specific dictionary learning is used to reduce computational time. This method is evaluated over two benchmark datasets such as ICL and Swedish leaf[2] datasets. Grayscale images of standard datasets are processed.

Munisami *et al.*[3] collected a dataset named 'Folio' consisting of 32 classes. The authors preprocessed the leaf images using rotation, gray-scaling, Otsu's thresholding, opening operation, inverse thresholding, edge extraction and edge filtering. After preprocessing, feature extraction methods such as convex hull, morphological information such as length, the width of the leaf, distance maps (horizontal, vertical, centroid radial maps) and color histogram are applied. Further, K-nearest neighbor (KNN) algorithm is used for classification.

Wu *et al.*[4] proposed a probabilistic neural network (PNN) for classification. The leaf images are converted from RGB image to binary image and boundary enhancement using Laplacian operator is applied. Twelve morphological features are used to extract information from the preprocessed leaf images. Principal Component Analysis (PCA) is applied to extract five salient features after feature extraction. These five principal variables act as the input vector to PNN.

Chaki *et al.*[5] extracted texture features using a Gabor filter and grey level co-occurrence matrix (GLCM). Shape features are extracted using curvelet transform and invariant moments. The efficiency of the author's proposed methods is analyzed with two classifiers namely, the neuro-fuzzy controller (NFC) and feedforward backpropagation multi-layered perceptron (MLP).

Bosilj *et al.*[6] proposed a pixel-based crop and weed discrimination approach using attribute profiles (AP). These AP contains morphological pixel descriptors and is compared with local binary pattern (LBP) and histogram of gradients (HoG) descriptors. Random forest classifier with 10 trees is used for classification. Two datasets used for evaluation are Sugar beets 2016 and Carrots 2017 dataset.

Tan *et al.*[7] proposed D-Leaf with three convolutional layers and three fully connected layers. D-leaf is used for feature extraction and hence, its final softmax layer is removed. Machine learning classifiers such as Support Vector Machine (SVM), Artificial Neural Network (ANN), k-Nearest-Neighbour (k-NN), Naïve-Bayes (NB) and CNN. Their method has been compared to AlexNet and fine-tuned AlexNet. MalayaKew, Flavia[4] and Swedish leaf[2] datasets are used for validating the model. The authors conclude that CNN is better to be used as feature extraction than for classification. Their proposed model of CNN is not completely being used and it is noticed that even fine-tuned AlexNet model performs better in comparison to D-Leaf.

Grinblat *et al.*[8] concentrated on leaf vein patterns for plant recognition using a deep convolutional neural network. The authors have pointed out that CNNs eradicated the need for handcrafted feature extraction methods. The deep convolutional neural network produced good accuracies for increased depth and is tested over three legume species (white bean, red bean and soybean).

Zhu *et al.*[9] combined the VGG 19 CNN model with SVM for plant identification using PlantCLEF dataset containing 1000 categories. The authors have compared their method with and without applying ReLU and softmax activation in the fully connected layers. Since, PlantCLEF contain plant organ oriented images (flower, leaf, fruit, branch and stem), their accuracies are calculated individually.

Atabay *et al.*[10] created its own five layered CNN network. Their CNN model contains two sets of convolution with a max-pooling layer each having 5 × 5 filters followed by 5 × 5, 1 × 1 convolutions and fully connected layers. The authors have used softmax in the fully connected layer. Hidden layers are analyzed on varying their activation functions as Rectified Linear Unit (ReLU) and Exponential Linear Unit (ELU). The authors observed though usage of ELUs produced good accuracies than ReLUs, ELUs consumed additional training time. This CNN architecture is tested with two standard datasets Flavia[4] (with ELUs) and Swedish datasets[2] (with ELUs). Horizontal reflection augmentation of datasets is considered for data augmentation. Shifting and rotation type of data augmentations are not carried out by the authors as the performance is too good with horizontal reflection augmentation. Grayscale images are used for this methodology. Usually, direct RGB images are given as inputs to CNN for image classification, except for resizing the images.

Jeon *et al.*[11] proposed a modified GoogleNet architecture for plant classification. GoogleNet architecture contains a set of convolution, pooling with three, five and two Inception blocks followed by pooling and fully connected layer. Modified GoogleNet architecture uses five inception blocks instead of three inception blocks in GoogleNet. The authors have used an image size of 229 × 229 × 3 and Flavia dataset is used for testing purpose. The authors have considered the eight shapes of leaves in Flavia dataset instead of their classes. Horizontally rotated and discoloured or damaged leaf images are also considered in the dataset. One drawback is that GoogleNet architecture consumes more than 8 hours and modified GoogleNet architecture takes an additional one hour. Though accuracy increases, time consumption also increases. This method might not be suitable for real-time prediction.

From the literature survey, these are some of the simple observations made. CNN models outperform the traditional way of classification. Traditional methods of classification are carried after applying a lot of preprocessing and feature extraction methods. Most of the CNN based methods in plant recognition use ReLU as activation function and stochastic gradient descent (SGD) for training. Overfitting is reduced by either data augmentation, batch normalization or dropout regularization methods. Many researchers working on plant recognition have concentrated on newer deep learning methods like AlexNet,[7] VGG-19,[9] own CNN,[8] GoogleNet,[10] transfer learning of CNN.[7] For plant classification based on leaves, LeNet-5,[12] the first CNN model has not yet been reported by researchers. Hence, a modified LeNet-5 deep learning method is proposed to recognize plant species. Adam optimizer is adapted for training the simple deep learning model. Our paper is organized as methodology, results and discussion followed by a conclusion.

2. METHODOLOGY

In deep learning, extensively CNN[13] is used for image classification type of applications. CNN contains a set of convolution layers with max pooling and fully connected layers. Convolution layer performs convolution in an image through sliding the window with weight matrix known as the kernel. Convolution layer contains characteristics like parameter sharing and sparse connections. In simple words, convolution in CNN is a set a multiplication and addition. Max pooling layer reduces the dimension of the image according to the window size of the kernel that is used. Fully connected layer unrolls 2D features into 1D features for categorizing the data with the help of neurons. Commonly, ReLU is the activation function used in hidden layers and softmax activation in the final fully connected layer.

LeNet-5[12] is the first Convolutional Neural Network (CNN) model and especially, it is developed for recognizing grayscale images of MNIST dataset. LeCun *et al.* created the CNN model with two sets of convolution layer (C1, C3) with subsampling layer (S2, S4), fully connected convolution (C5) is present with 120 feature maps and two fully connected layers (F6, output) are used. So, there are seven layers in LeNet-5. Subsampling layer is the average pooling. The first convolution has six 5×5 filters with stride, $s = 1$ and padding are not used for convolutions. Average pooling with 2 × 2 filter and stride 2 follows the convolution. The second convolution has sixteen 5 × 5 filters and is followed by an average pooling layer with a 2 × 2 filter using a stride, s = 2. Except for the final output layer, it uses tanh activation function. Final layer uses softmax activation function.

2.1 Architecture

The flow diagram for our plant recognition system is shown Figure 1. Images are resized to 100 × 100 × 3 using linear interpolation method. All pixels of the images are normalized. These normalized pixel intensities are forwarded as input to the deep networks. Normalizing of pixel intensities (each pixel divided by 255) is the preprocessing step. We are using six layers for our modified LeNet-5 architecture.

Rectified linear unit (ReLU) activation function and a max-pooling layer follow two convolution layers with a 5 × 5 filter. The outputs of the second max-pooling layer are converted from 2D vectors into 1D vectors. These 1D vectors are inputs to the first fully connected layer.

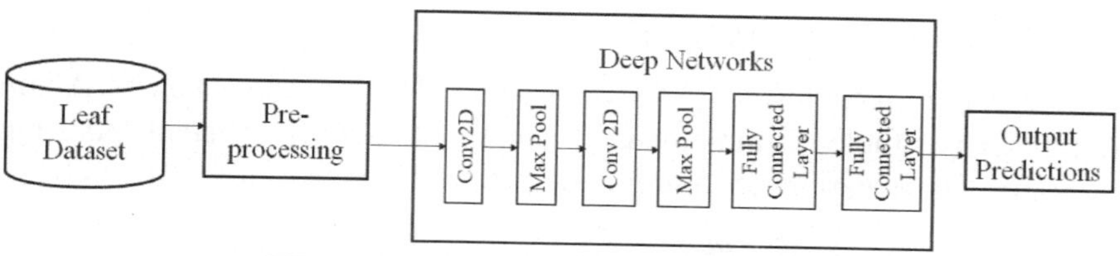

Fig. 1: The Flow of the Proposed Method

Table 1: Configuration of the Proposed Method [Input shape (100 × 100 × 3)]

Layer	*Filter*	*No. of Filters*	*Stride*	*Activation Function*	*No. of Parameters*	*Output Shape*
Convolution	5 × 5	20	S = (1, 1)	ReLU	1520	(100, 100, 20)
Max Pooling	2 × 2	–	S = (2, 2)	–	–	(50, 50, 20)
Convolution	5 × 5	50	S = (1, 1)	ReLU	25050	(50, 50, 50)
Max Pooling	2 × 2	–	S = (2, 2)	–	–	(25, 25, 50)
Fully connected	–	–	–	ReLU	15625500	(None, 500)
Fully connected	–	–	–	Softmax	16032	Num_classes(32)

Two fully connected layers are used. First fully connected layer has ReLU as the activation function and it consists of 500 neurons. The outputs are predicted based on the second fully connected layer with softmax activation function. The configurations used for our work is mentioned in Table 1. We use 'same' padding for modified LeNet-5 in order to preserve the information.

Equations of ReLU and softmax activation functions are shown in (1) and (2). In equation (1), input, x is always greater than 0 and maximum among 0 and y is the output of ReLU function, y. Softmax activation function, $S(y_i)$ provides an output of probabilities between 0 and 1 for each class s_i varying from j = 1 to C classes. We use categorical cross-entropy loss function to determine the error between the predicted and actual output. This categorical cross-entropy loss (CCE) function is calculated for target labels, t_i varying from i = 1 to C, number of classes and is given in equation (3).

$$y = \max(0, x) \quad \ldots (1)$$

$$S(y_i) = \frac{e^{s_i}}{\sum_j^C e^{s_j}} \quad \ldots (2)$$

$$CCE = -\sum_i^C t_i \log(S(y_i)) \quad \ldots (3)$$

2.2 Training Deep Networks

The learning algorithm for Adam optimizer[14–15] is written below. Adam optimizer is simple to implement. It is computationally efficient requiring very little hyperparameter tuning. Adam weight update rule is given by,

$$\theta_{t+1} = \theta_t - \frac{\eta}{\sqrt{\hat{v}_t + \epsilon}} \hat{m}_t \quad \ldots (4)$$

Where, θ_t, old weight,

θ_{t+1}, new weight,

η, learning rate,

$\hat{m}_t$, the first moment based on decay rate, β_1,

$\hat{v}_t$, the second moment based on decay rate, β_2,

ϵ, is a small number that prevents division by zero.

3. RESULTS AND DISCUSSION

The result analysis carried out with a benchmark leaf dataset, Flavia. There are 32 classes in Flavia dataset and 50 images are present in each class. Sample images of Flavia dataset are shown in Figure 2. The dataset is split into 70% for training and 30% for testing. The deep networks are trained using Adam optimizer and compared with other optimizers. We analyze and tune the hyperparameters, namely, learning rate, batch size and epochs. For our analysis over Flavia dataset, the results are recorded when the loss curves are decreasing and accuracy curves are increasing such that the data does not overfit the model. The experimental results are taken using NVIDIA Titan X GPU along with i7 processor CPU and we use deep learning framework, keras.

3.1 Analysis of Optimizers

For our proposed deep networks, we vary the optimizers such as SGD, Adam, Adamax, Adadelta by setting the values of epoch and batch size as 50 and 32, respectively. We noticed that Adam optimizer showed a higher accuracy of 88.54% Figure 3 shows the top-1 and top-5 accuracies of various optimizers. For further analysis, we used Adam optimizer.

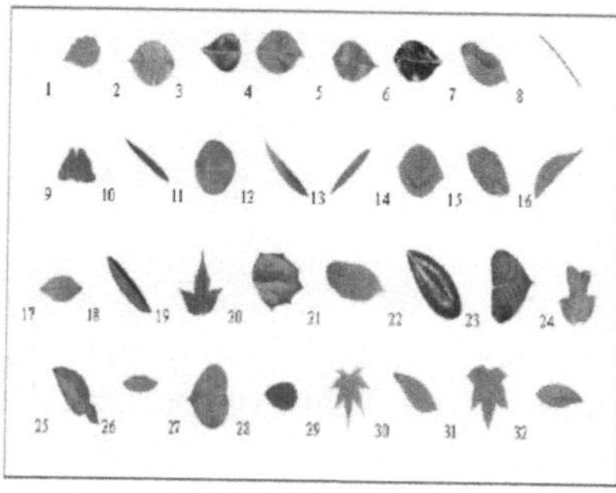

Labels	Botanical Name	Labels	Botanical Name
1	*Koelreuteria paniculata Laxm.*	17	*Prunus persica (L.) Batsch*
2	*Ilex macrocarpa Oliv.*	18	*Manglietia fordiana Oliv.*
3	*Pittosporum tobira (Thunb.) Ait. f.*	19	*Acer buergerianum Miq.*
4	*Chimonanthus praecox L.*	20	*Mahonia bealei (Fortune) Carr.*
5	*Cinnamomum camphora (L.) J. Presl*	21	*Aesculus chinensis*
6	*Viburnum awabuki K.Koch*	22	*Magnolia grandiflora L.*
7	*Osmanthus fragrans Lour.*	23	*Populus canadensis Moench*
8	*Cedrus deodara (Roxb.) G. Don*	24	*Liriodendron chinense (Hemsl.) Sarg.*
9	*Ginkgo biloba L.*	25	*Citrus reticulata Blanco*
10	*Phyllostachys edulis (Carr.) Houz.*	26	*Berberis anhweiensis Ahrendt*
11	*Lagerstroemia indica (L.) Pers.*	27	*Cercis chinensis*
12	*Nerium oleander L.*	28	*Indigofera tinctoria L.*
13	*Podocarpus macrophyllus (Thunb.) Sweet*	29	*Acer Palmatum*
14	*Prunus serrulata Lindl. var. lannesiana auct.*	30	*Phoebe nanmu (Oliv.) Gamble*
15	*Ligustrum lucidum Ait. f.*	31	*Kalopanax septemlobus (Thunb. ex A.Murr.) Koidz.*
16	*Tonna sinensis M. Roem.*	32	*Cinnamomum japonicum Sieb.*

Fig. 2: Sample Images and Labels of Flavia Dataset

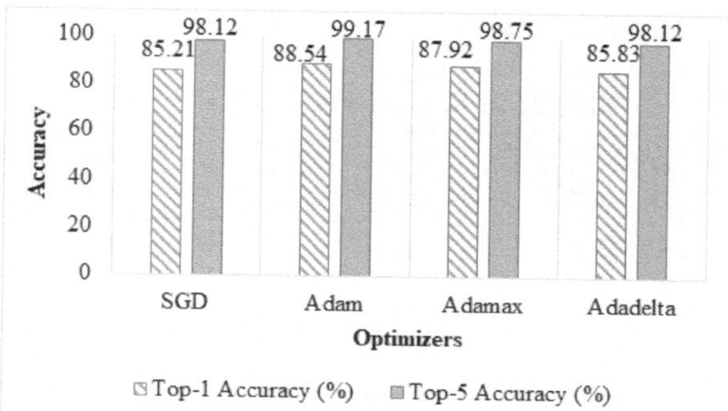

Fig. 3: Accuracies of Various Optimizers

3.2 Effect of Learning Rate

We modified the learning rates for Adam optimizer at a fixed number of epochs, 50 and batch size, 32. These results are shown in Figure 4. A slight decrease in accuracy is recorded for the learning rate, 0.0001 compared to other learning rates. Best accuracy of 89.17% is noted for learning rate, 0.0001. When we further increased (e.g. 0.1, 0.01) or decreased (e.g. 0.000001), we obtained poorer accuracies. Figure 4 shows the top-1 and top-5 accuracies on reducing the learning rate (divided by 10) from 0.001 to 0.00001.

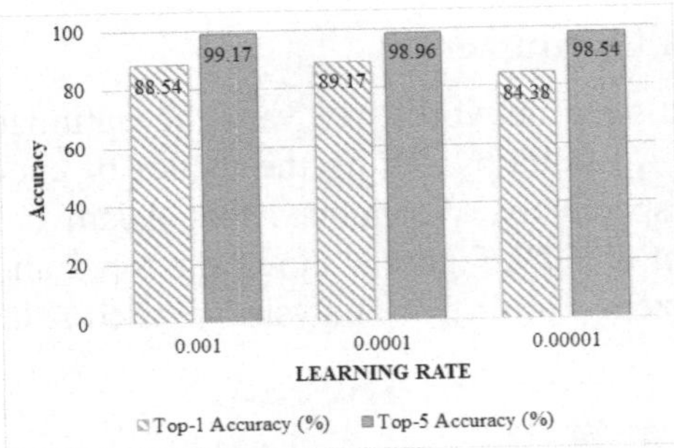

Fig. 4: Accuracies of Adam Optimizer

3.3 Effect of Batch Size

We altered the batch sizes for training the model as 1, 8, 16, 32 and are shown in Table 2. The results of batch size are taken by setting learning rate and a number of epochs as 0.0001 and 50 for Adam optimizer. We observed that as the batch size is increased, accuracy increases and training time decreases. The training time taken for batch size 32 is 63 s. We have taken batch size as 32 for the analysis on varying epochs as it took less time compared to other batch sizes.

Table 2: Results on Varying Batch Size

Batch Size	*Top-1 Accuracy (%)*	*Top-5 Accuracy (%)*	*Loss*	*Training Time(s)*	*Testing Time(s)*
1	87.92	97.92	0.98	829.54	0.54
8	88.54	98.54	0.52	143.1	0.53
16	89.58	98.96	0.43	93.16	0.6
32	89.17	98.96	0.37	63	0.62

3.4 Effect of Epochs

We increased the number of epochs from 50 to 500 for Adam optimizer with learning rate, 0.0001. The model stabilized after 100 epoch. When the number of epochs is 200, 300, 400, 500, the loss kept increasing and accuracy did not improve. Thus, we are concluding that for 100 epochs are enough to train the model and we obtained an accuracy of 89.58%.

Figure 5 (a) and (b) shows the accuracy and loss plot for the deep networks observed till 100 epochs for Adam optimizer with learning rate, 0.0001. Precision, recall and f1-scores for 100 epochs trained using Adam optimizer with learning rate, 0.0001 are 0.9, 0.9 and 0.9.

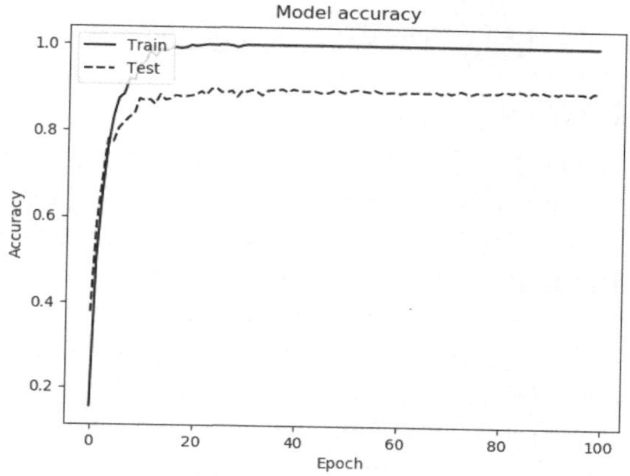

(a)

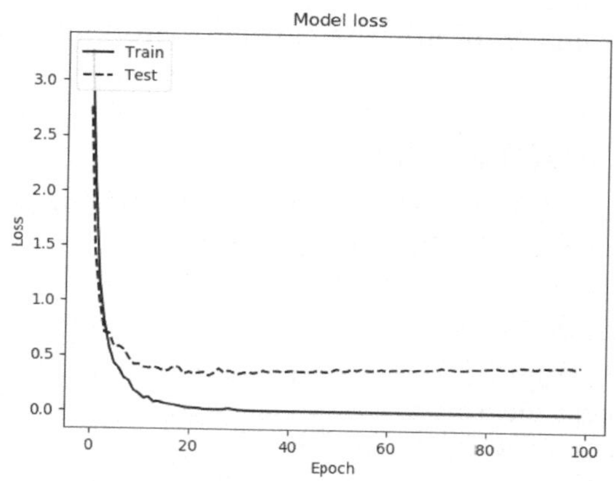

(b)

Fig. 5: (a) Accuracy Curves for Training and Testing.
(b) Loss Curves for Training and Testing.

3.5 Evaluation of Benchmark Dataset

Flavia dataset is a widely used leaf dataset for plant recognition. Table 3 shows the validation of our proposed work over other existing methodologies. Our proposed method outperforms classifiers such as k-Nearest Neighbor (KNN), Multi-layered Perceptron (MLP) and customized CNN.

Table 3: Existing Works on Flavia Dataset

Method	*Accuracy*
Customized CNN[16]	49.8%
KNN[3]	86%
MLP[4]	85.6%
Our Method	89.58%

4. CONCLUSION

In our proposed work, plant recognition task has been carried out using the modified version of the first deep network model, LeNet-5. We used the same size of the filter (5 × 5) and strides as in LeNet-5 architecture and made use of ReLU activation function instead of tanh. This method is tested over standard leaf dataset, Flavia containing 32 plant species each of 50 images. We observed that the architecture stabilized after training our images with 100 epochs and Adam optimizer with learning rate, 0.0001 with an accuracy of 89.58%. Even though the number of layers in the modified LeNet-5 architecture look smaller, the number of parameters utilized is about 15,668,102. The number of parameters in our proposed work is lesser than VGG models that use 138 million parameters. As future work, a simpler, deeper and parameter efficient CNN architecture will be developed.

ACKNOWLEDGEMENT

The authors would like to thank NVIDIA for providing NVIDIA Titan X GPU under the University Research Programme.

REFERENCES

[1] Zhang, S., Zhang, C., Wang, Z. and Kong, W. (2018). Combining sparse representation and singular value decomposition for plant recognition. *Applied Soft Computing*, 67, pp. 164–171.

[2] Söderkvist, O. (2001). Computer vision classification of leaves from swedish trees.

[3] Munisami, T., Ramsurn, M., Kishnah, S. and Pudaruth, S. (2015). Plant leaf recognition using shape features and colour histogram with K-nearest neighbour classifiers. *Procedia Computer Science*, 58, pp. 740–747.

[4] Chaki, J., Parekh, R. and Bhattacharya, S. (2015). Plant leaf recognition using texture and shape features with neural classifiers. *Pattern Recognition Letters*, 58, 61–68.

[5] Wu, G., Bao, F.S., Xu, E.Y., Wang, Y.X., Chang, Y.F. and Xiang, Q.L. (2007), December. A leaf recognition algorithm for plant classification using probabilistic neural network. *In 2007 IEEE international symposium on signal processing and information technology*, pp. 11–16.

[6] Bosilj, P., Duckett, T. and Cielniak, G. (2018). Analysis of morphology-based features for classification of crop and weeds in precision agriculture. *IEEE Robotics and Automation Letters,* 3(4), pp. 2950–2956.

[7] Tan, J.W., Chang, S.W., Kareem, S.B.A., Yap, H.J. and Yong, K.T. (2018). Deep Learning for Plant Species Classification using Leaf Vein Morphometric. IEEE/ACM transactions on computational biology and bioinformatics.

[8] Grinblat, G.L., Uzal, L.C., Larese, M.G. and Granitto, P.M. (2016). Deep learning for plant identification using vein morphological patterns. *Computers and Electronics in Agriculture,* 127, pp. 418–424.

[9] Zhu, H., Liu, Q., Qi, Y., Huang, X., Jiang, F. and Zhang, S. (2018). Plant identification based on very deep convolutional neural networks. *Multimedia Tools and Applications,* pp. 1–19.

[10] Atabay, H.A. (2016). A convolutional neural network with a new architecture applied on leaf classification. *IIOAB J,* 7(5), pp. 226–331.

[11] Jeon, W.S. and Rhee, S.Y. (2017). Plant leaf recognition using a convolution neural network. *International Journal of Fuzzy Logic and Intelligent Systems,* 17(1), pp. 26–34.

[12] LeCun, Y., Bottou, L., Bengio, Y. and Haffner, P. (1998). Gradient-based learning applied to document recognition. *Proceedings of the IEEE,* 86(11), pp. 2278–2324.

[13] Goodfellow, I., Bengio, Y. and Courville, A. (2016). Deep learning. MIT press.

[14] Kingma, D.P. and Ba, J. (2014). Adam: A method for stochastic optimization. arXiv preprint arXiv:1412.6980.

[15] Ruder, S. (2016). An overview of gradient descent optimization algorithms. arXiv preprint arXiv:1609.04747.

[16] Liu, Albert and Huang, Yangming, http://cs229.stanford.edu/proj2016/report/LiuHuang-PlantLeafRecognition-report.pdf

Image Contrast Enhancement Using Xilinx System Generator

V. Vigneshwarar[1] and S.P. Joy Vasantha Rani[2]
Department of Electronics Engineering, MIT Campus, Anna University
E-mail: [1]vigwar72@gmail.com; [2]joy_mit@annauniv.edu

ABSTRACT: *This paper presents a contrast enhancement method (CA). It is a method that is used to enhance images for viewing process or for further analysis of images. Main idea behind contrast enhancement techniques is to increase contrast and to preserve original brightness of images. In this paper, a Modified Adaptive Gamma Correction with Weighting Distribution algorithm is proposed for contrast enhancement. This novel hardware-oriented contrast enhancement algorithm is modelled in Xilinx System Generator. The modified algorithm has PSNR value 25% higher than AGCWD, indicating higher image quality. The proposed technique is basically an improvement over AGCWD technique preservation than MAGCWD technique.*

Keywords: Histogram Equalization, Contrast Enhancement, Gamma Correction, Peak Signal to Noise Ratio, Root Mean Square Error.

1. INTRODUCTION

Contrast enhancement[2] refers to the image enhancement by adjusting the dynamic range of pixel intensity and also plays the improvement of visual quality for pattern recognition and digital image processing. It improves the perceptibility of objects[3] in the scene by enhancing the brightness difference between objects and their backgrounds. It improves the brightness differences uniformly across the dynamic range of the image, and differences in the shadow (dark), greys, highlight (bright) regions at the expense of the brightness differences in the other regions.

Figure 1 shows the block diagram of contrast enhancement. The histogram is calculated using Image Statistics Computation (ISC) for grey scale and color images. It is a graphical representation of image. It shows the frequency of pixel intensity values. Weighted Probability Density Function (WPDF) is calculated from the histogram and the Smoothed Cumulative Distribution function (SCDF) is used for calculating the cumulative density function (CDF). Adaptive Gamma Correction (AGC) is mainly used to control the overall brightness of the image. Gamma can be any value between 0 and 1. If gamma is 1, the mapping is linear. If gamma is less than 1, the mapping is weighted towards higher output values. If gamma is greater than one, mapping is weighted toward lower output values. Finally, Luminance Transformation is used for correcting the intensity of the pixel depending upon the gamma value of the brightness of the image.

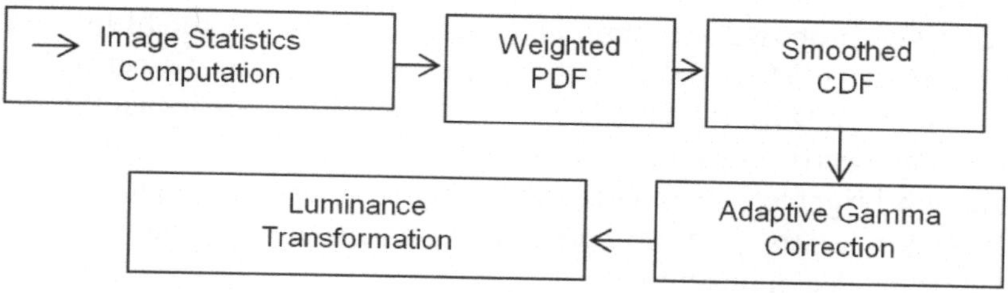

Fig. 1: Block Diagram for Contrast Enhancement

2. RELATED WORK

Kim and Chung (2008), the image is separated areas tonal enhancements improve the brightness depending up on the basis of mean of input image. It implies that in this technique of decomposing the input image only once, it is decomposed recursively up to a recursion level r, therefore 2r sub images that can be generated. Each sub-images are equalized independently with histogram equalization method. If r = 0, that means no sub-image decomposition is done, i.e. it is equivalent to HE method. If r = 1, that it is equivalent to BBHE.[6] The level of brightness preservation will increase with the increase of number of recursive mean separations. Though it is recursive in nature, RMSHE also allows scalable brightness preservation. Wan, Chen, and Zhang, proposed the original image is divided into two equal area sub-images based on gray level probability density function of input image. The DSIHE technique for contrast enhancement decomposes an image into two equal area sub-images, one dark and one bright, following the equal area property. Resulting image of dualistic sub-image histogram equalization (DSIHE) is obtained after the two equalized sub-images will be composed into one image. This is similar to BBHE except difference is that in this method DSIHE chooses to separate the Histogram based on its mean gray scale images. It not only enhance image information effectively but also keeps Input luminance well enough that has to be enhanced effectively. Kim, proposed a bi-histogram preserving method, it produces an output image with the value of the brightness (the mean grey-level) located in the middle of the mean of the input image and the median grey-level. One of the sub images is set of samples a less than or equal to the mean whereas the other one is the set of samples greater than the mean. Separation intensity is used in order to construct the image respectively. The output mean brightness of the BBHE method is a function of the input mean brightness. W.C. Chen, S.C. Huang, and T.Y. Lee,[5] it follows the basic principle of the BBHE and DSIHE methods of decomposing an image before applying the histogram Equalization method to equalize the resulting sub-images independently, W.C. Chen proposed the minimum mean brightness error bi-histogram equalization (MMBEBHE) method. The main difference between the BBHE and the DSIHE methods and the MMBEBHE one is that threshold level that decomposes the image into two sub-images such that the minimum brightness difference between the input and output images is achieved, whereas the former methods consider only the input image to perform the decomposition.

3. MODIFIED AGCWD ALGORITHM

The main objective of the proposed technique is to transform an image into good quality images compared to AGCWD. This is done by increasing the contrast and brightness and give clear output image. To achieve this, we propose an adaptive gamma correction and along with Brightness preserving histogram equalization method which determines an intensity transformation function according to the characteristics of the input image.

Figure 2 shows the algorithm flow of modified AGCWD. The algorithm accepts color image, grey scale and real time images. A dataset of 10

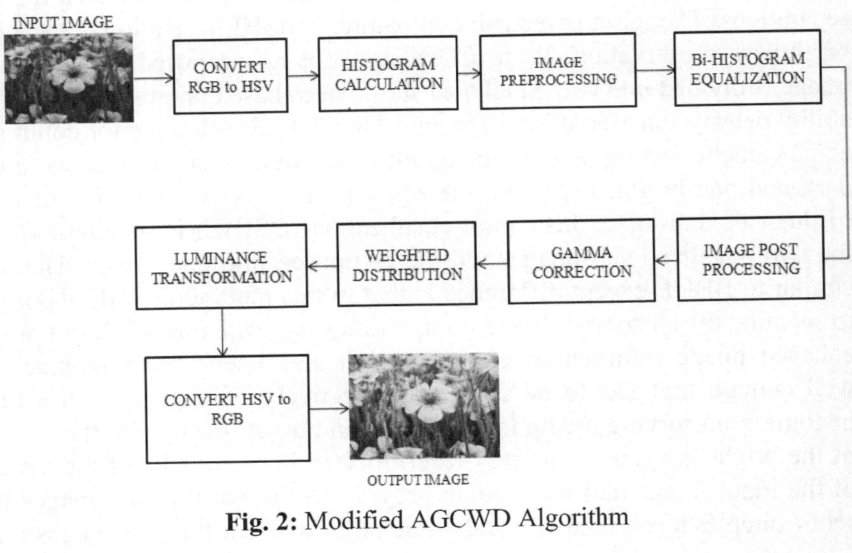

Fig. 2: Modified AGCWD Algorithm

standard images has been taken, from which five are grey scale and five are colored images. To convert RGB to HSV, we need a color conversion component by using that component we can convert color information between 6 different color spaces and separate each components into Hue, saturation, value respectively.

$$Y = 0.3R + 0.59G + 0.11B \quad \text{... (1)}$$

3.1 Histogram Calculation

Let X = {X(i, j)} denote a given image composed of Ldiscrete gray levels denoted as $\{X_0, X_1, \ldots\ldots, X_{L-1}\}$ where X(i, j) represents an intensity of the image at the spatial location (i, j) and X(i, j) £$\{X_0, X_1, \ldots, X_{L-1}\}$. F or a given image X, the probability density function p (X_k) is defined as,

$$P(X_k) = n^k/n \quad \text{... (2)}$$

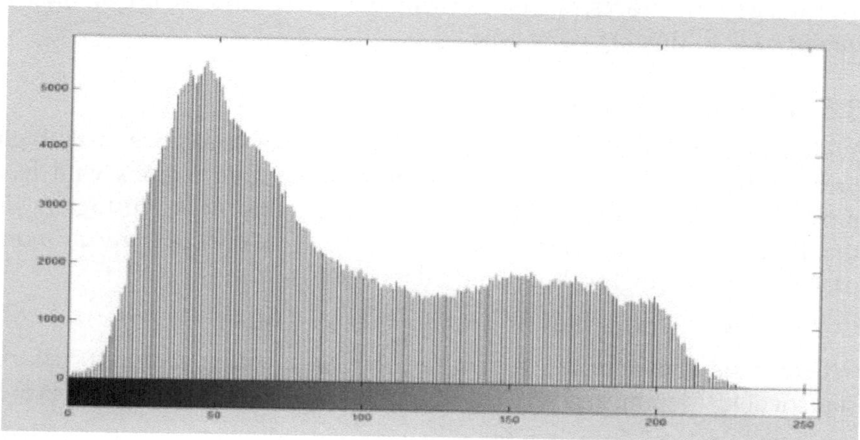

Fig. 3: Histogram Calculation

Fork = 0, 1, ..., L–1. Where n_k represents the number of times that the level X_K appears in the input image X and n is the total number of samples in the input image. The relative frequency (proportion) of each class, represents frequency of pixel intensity values, is calculated by,

Relative frequency = Class frequency/total no. of the samples

3.2 Preprocessing

It is used to convert two dimensional image into one dimensional image. For Preprocessing it uses a buffer block which converts frame input to scalar samples at higher sampling rate.

3.3 Bi-Histogram Equalization

The BBHE is an extension of a typical histogram equalization, which utilizes independent histogram equalizations over two sub images obtained by decomposing the input image based on its mean. The ultimate goal behind the BBHE is to preserve the mean brightness of a given image while enhancing the contrast of a given image. It not only enhance image information effectively but also keep the original image Input luminance well enough to be used and gives better quality when compared to HE. The separation intensity XT is presented by the input mean brightness value, which is the average intensity of all pixels that construct the input image. After this separation process, these two histograms are independently equalized by Histogram Equalization.

3.4 Post Processing

The image post processing blocks which are used to convert the image output back to floating point type.

3.5 Weighted Distribution

Weighted distribution is applied so that the image regions with high probability should not get over enhanced and image regions with less probability should not be less enhanced. In this input histogram is modified in the way that less frequent gray levels are given more probabilities or weights. Weighted Pdf is calculated as,

$$Pdf_w\,(l) = Pdf_{max}\,(Pdf\,(l) - Pdf_{min}/Pdf_{max} - Pdf_{min})^{\alpha} \quad \ldots (3)$$

Where α is the adjusted parameter, pdf_{max} is the maximum pdf of statistical histogram, and pdf_{min} is minimum pdf. Now modified cdf is as,

$$Cdf_w\,(l) = \sum pdf_w(l)/\sum Pdf_w \quad \ldots (4)$$

3.6 Gamma Correction

After weighted distribution the sub-histograms are mapped to final histogram and gamma correction is applied. In this technique gamma correction is applied using normalized cumulative density function (cdf). Gamma correction is done as:

$$T\,(l) = lmax\,(l/lmax)\,\gamma = lmax\,(l/lmax)\;1 - Cdf\,(l) \quad \ldots (5)$$

Gamma can be calculated as,

$$\gamma = 1 - Cdf_w\,(l) \quad \ldots (6)$$

Gamma can be any value between 0 and 1, if gamma is 1, the mapping is linear. If gamma is less than 1, the mapping is weighted towards higher output values. If gamma is greater than one, mapping is weighted towards lower output values. It produces comparatively good results than previous algorithm (AGCWD) and increases the brightness of the input image and gives desired visual quality and color contrast. The difference between the AGCWD and MAGCWD are analyzed by using PSNR and RMSE.

4. PERFORMANCE ANALYSIS

In this section, we demonstrate the performance of image quality for the proposed method Modified AGCWD in comparison with AGCWD methods including BBHE and RMSHE. As test images, we selected 4 images: *Flower*, *House*, *Elephant*, *Leaf*, which have been previously used in many other methods to show their performances of *Root Mean Square Error (MSE)* and *Peak Signal to Noise Ratio (PSNR).*[1,3,4,7] *MSE* can be obtained by using matlab function by comparing original image and enhanced image. This parameter is mainly used to measure the change in original brightness of image. *PSNR* can be obtained by a difference between the original image of the signal noise and the enhanced image of the noise. *PSNR* is given by,

$$PSNR = 20 \log [g (p-1)] - 10\log\sqrt{MSE} \quad \ldots (7)$$

Where, p - no of pixel intensity, g - Maximum number of gray levels.

4.1 System Generator Model for Modified AGCWD

Figure 4 show that System generator for Modified AGCWD works with the Simulink model based methodology. First image is taken, next step convert RGB to HSV by using color space conversion, next step to

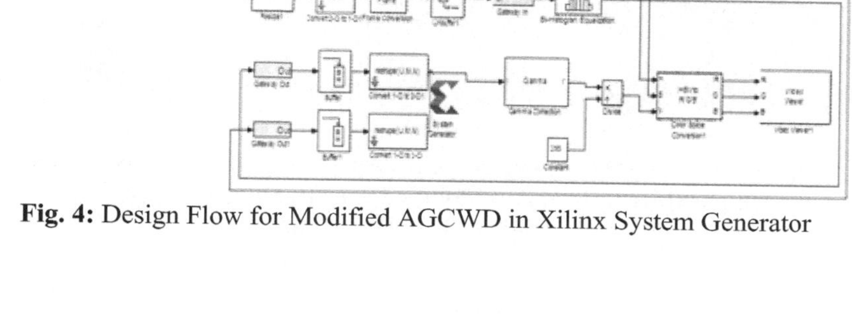

Fig. 4: Design Flow for Modified AGCWD in Xilinx System Generator

separate R|G|B color format, then equalize the image by using Bi-Histogram equalization technique after that gamma correction done by calculating probability density function and cumulative density function depending upon the values obtained in the BBHE. After that convert HSV to RGB in order to obtain the color image and output gives the desired brightness when compared to AGCWD.

The performance of image quality for the proposed method Modified AGCWD in comparison with AGCWD methods. For visual assessment initially two color images flower shown in Figure 5, Elephant shown in Figure 7, one gray scale image shown in Figure 6. The Modified AGCWD gives good visual quality images when compared to AGCWD.

(a) (b) (c)

Fig. 5: Flower (a) Original Image (b) Adaptive Gamma Correction with Weighted Distribution (c) Modified Adaptive Gamma Correction with Weighted Distribution

(a) (b) (c)

Fig. 6: House (a) Original Image (b) Adaptive Gamma Correction with Weighted Distribution (c) Modified Adaptive Gamma Correction with Weighted Distribution

(a) (b) (c)

Fig. 7: Elephant (a) Original Image (b) Adaptive Gamma Correction with Weighted Distribution (c) Modified Adaptive Gamma Correction with Weighted Distribution

5. GRAPHICAL ANALYSIS

Table 1 shows that Peak Signal to Noise ratio and Root Mean Square Error for Modified AGCWD and AGCWD. The PSNR value for AGCWD and Modified AGCWD on dataset of grayscale images that are shown in Table 1. The Modified AGCWD gives higher PSNR rate than the existing AGCWD method indicating the greater the PSNR value better will be the image quality. The brightness preservation can be estimated from Root Mean Square error parameter. Results of Modified AGCWD technique are better from AGCWD technique as the proposed technique Modified AGCWD.

The contrast enhancement can be observed by PSNR value, The PSNR value for AGCWD and Modified AGCWD for the dataset of gray scale images show in Figure 8 greater the PSNR value better will be the image quality. The Modified AGCWD for Grayscale images gives higher PSNR rate giving high quality results when compared to a AGCWD. The Brightness preservation can be estimated from RMSE. Root mean square values for AGCWD and Modified AGCWD for the dataset of grey color images are shown in Figure 9 result shows that brightness is preserved in

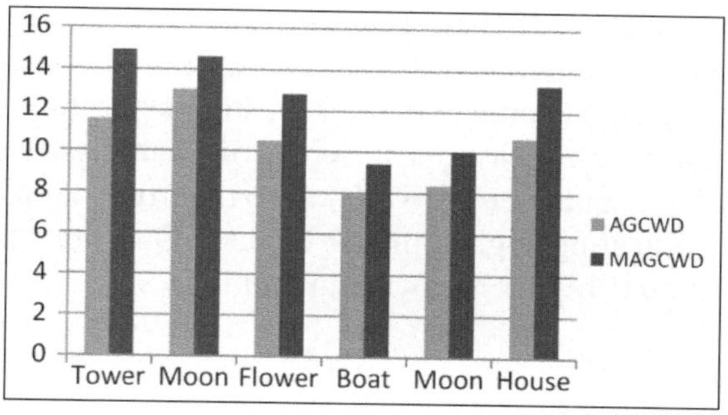

Fig. 8: Evaluation for PSNR

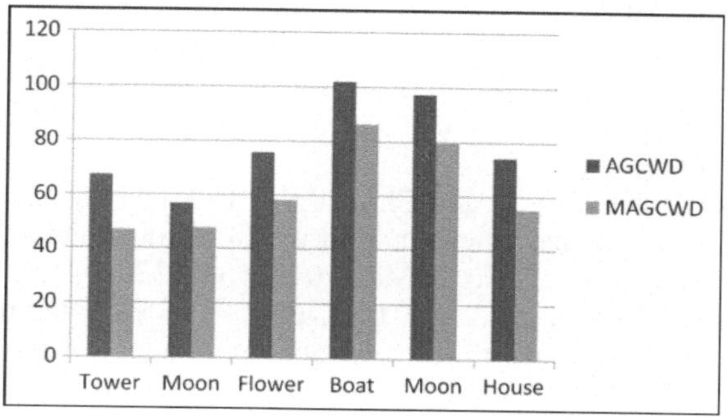

Fig. 9: Evaluation for RMSE

Table 1: Performance Analysis for PSNR and RMSE-Gray Scale Images

S. No.	*Gray Scale Images*	*Pixel Range (height & width)*	*AGCWD-PSNR (dB)*	*MAGCWD-PSNR (dB)*	*AGCWD-RMSE*	*MAGCWD-RMSE*
1.	Tower	1498×968	11.58	14.91	67.18	46.87
2.	Moon	1280×800	13.04	14.58	56.76	47.58
3.	Flower	640×1280	10.52	12.83	75.87	58.21
4.	Boat	1000×1500	7.96	9.42	101.97	86.19
5.	Moon	1280×804	8.37	10.05	97.22	80.16
6.	House	728×546	10.71	13.30	74.27	55.13

proposed technique Modified Adaptive Gamma correction with weighted distribution when compared to Adaptive Gamma correction with weighted distribution (AGCWD).

6. CONCLUSION

In this paper, contrast enhancement algorithm is used. The bi-histogram equalization method (BBHE) is combined with the existing AGCWD method in order to enhance the image. In this method, the mean brightness and contrast are improved. The images are qualitatively and quantitatively analyzed and the results show that the proposed method Modified AGCWD performs well compared to the existing algorithm AGCWD. The quantitative results are measured in terms of Peak signal to noise ratio (PSNR) and Mean square error (MSE). The modified algorithm has high PSNR value, indicating higher image quality. One drawback observed with the algorithm is higher hardware utilization.

REFERENCES

[1] Wilson, Garnet, "FPGA Implementation of Hardware Efficient Algorithm for Image Contrast Enhancement Using Xilinx System Generator", *Procedia Technology,* Vol. 24, 2016, pp. 1141–1148.

[2] Chenand, W.C. and Huang, S.C., "A New Hardware-Efficient Algorithm and Reconfigurable Architecture for Image Contrast Enhancement", *IEEE Trans. Image Process.*, Vol. 23, No. 10, pp. 1057–7149, Oct. 2014, pp. 1741–1747.

[3] Huang, S.C., Cheng, F.C. and Chiu, Y.S., "Efficient contrast enhancement using adaptive gamma correction with weighting distribution", *IEEE Trans. Image Process,* Vol. 22, No. 3, Mar. 2013, pp. 1032–1041.

[4] Chen, W.C., Huang, S.C. and Lee, T.Y., "An efficient re-configurable architecture design and implementation of image contrast enhancement algorithm", in *Proc. IEEE 9th Int. Conf. High Perform. Computer Communication,* Jun. 2012, pp. 1741–1747.

[5] Chiu, Y.S., Cheng, F.C. and Huang, S.C., "Efficient contrast enhancement using adaptive gamma correction and cumulative intensity distribution", in *Proc. IEEE Int. Conf. Syst., Man, Cybern*, Oct. 2011, pp. 2946–2950.
[6] Kim, M. and Chung, M., "Recursively separated and weighted histogram equalization for brightness preservation and contrast enhancement", *IEEE Trans. Consum. Electron.*, Vol. 54, No. 3, Aug. 2008, pp. 1389–1397.
[7] Wan, Y., Chen, Q. and Zhang, B., "Image enhancement based on equal area dualistic sub-image histogram equalization method", *IEEE Trans. Consum. Electron.*, Vol. 45, No. 1, Feb. 1999, pp. 68–75.
[8] Kim, Y.T., "Contrast enhancement using brightness preserving bi-histogram equalization", *IEEE Trans. Consum. Electron*, Vol. 43, No. 1, Feb. 1997, pp. 1–8.

Energy Efficiency Improvement in WSN through Software Defined Network

K. Tamil Kodi[1] **and S. Vasuhi**[2]

Department of Electronics Engineering, Madras Institute of Technology, Anna University, Chennai–600 044

E-mail: [1]tamilkodik@gmail.com; [2]vasuhi_s@annauniv.edu

ABSTRACT: *The smart technology in many areas, have gained more attention on Wireless Sensor Network (WSN). By using a software-defined network approach, energy consumption can be reduced effectively.*

Software-defined networking is an emerging model, which the vertical integration to provide the adaptability to this network. Thus, the implementation of Software Defined Network (SDN) on handling the energy consumption as well as the energy efficiency based on the recent work can be done using the NS2 simulator and pox controller. The algorithm like efficient routing algorithm selection is noted. In this paper, a framework for energy efficiency in SDN is presented. Software-defined networking is an open flow protocol rather than the traditional network. Open Flow protocol is a communication protocol in the environment of SDN which enables the controller to directly interact with the forwarding plane of sensor nodes. Routing in sensor nodes by the controller and communication between the switch and the controller via Open Flow protocol. This improves the efficiency of energy in the network.

Keywords: Software Defined Network (SDN), POX Controller, Energy Consumption, Performance, NS2, Traffic Management.

1. INTRODUCTION

The recent analysis in large scale digitization of data, calls significant changes in the communication technologies and network system platforms.[10] Software Defined Networking (SDN) is a platform that provides programmability, automation, and control of network functions by dissociating the control plane and the data plane of the network. It is important to optimize the energy consumption of network equipment and functions in SDN for improving the performance of WSN. Therefore, energy optimization has become a key challenge in performance evaluation of SDN. Unlike the traditional networks, where the entire system needs to be reconfigured to upgrade the system, only the software needs to be updated in the SDN, more convenient for upgrading and reduction of overall cost. SDN has gained attention as it is flexible and can be programmed. The best features of SDN is the hardware free design structure hence the cost can be highly reduced effectively with this network as shown in Figure 1. The structural design for the WSN will most probably include several limitations of the time is battery operated. This means that the available energy should be managed efficiently. SDN has the potential to facilitate the deployment and management of network applications and services with greater efficiency.

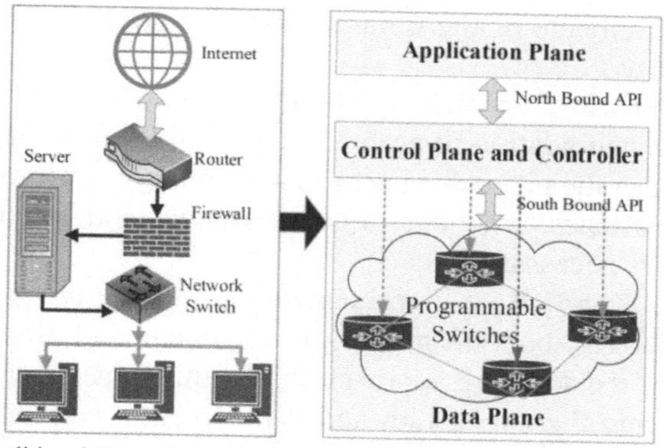

(a) Traditional Network Architecture (b) SDN Architecture

Fig. 1: Traditional and SDN Architecture

For improving the performance of SDN. It has been reported that there is an increase in energy consumption by computing and network devices as shown in Figure 2. The data centers to provide reliable and timely services.

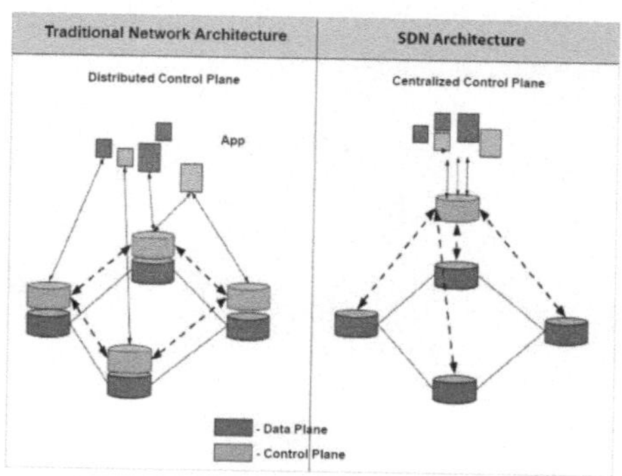

Fig. 2: The Control Structure of SDN Architecture

The network, as a crucial component of data center infrastructure, consumes a significant part of total energy (up to 20%).

The paper is organized as follows. The section 2 provides motivation and objective of SDN then we describe related works in section 3. We present corresponding schemes and algorithm in section 4. The implementation of an algorithm and report the results in section 5. Finally, the paper is concluded with a brief summary of contributions.

2. MOTIVATION AND OBJECTIVE

Today, many numbers of heterogeneous applications are the backbone network of any organization or publicly accessible network. On the other side, the requirements of the organizations are stringent regarding cost, QoS and security.

In such scenarios, the increase of traffic with varying requirements trivially cause degradation in performance of SDN controllers. To maintain and enhance the network's performance, energy consumption of SDN Controllers, switches, and links in heterogeneous and live networks with varying input is necessary.

Hlabishi I. Kobo *et al.,*[1] presented the survey on wireless sensor network challenges about the network management and lifetime of sensors by introducing the software approach in obtaining energy level consumption. Considering the application and deployment scenarios, as shown in Figure 3. The main objective of our work is to develop a novel optimized energy consumption framework for SDN and efficient routing algorithm selection for energy-aware traffic management and load balancing.

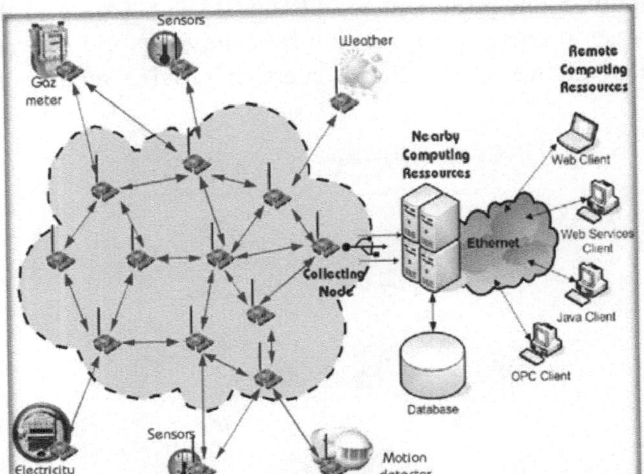

Fig. 3: The Application Scenario of WSN

3. RELATED WORKS

There have been a number of research works focused on optimizing energy in Software Defined Network. Most of these works contribute to different aspects to reduce energy consumption in SDN devices. The application based on their configuration of WSN[12] needs to be designed. Thus the IEEE configuration of WSN as shown in Table 1 for the specific application should be followed for the various applications.

Table 1: IEEE Configuration of WSN

IEEE 802.15.4	
Layers	Physical
Frequency	868/915 MHz 2.4 GHz
Data rate	0.25 Mbps
Range	10 to 100 m
Addressing	8 bit or 16 bit
Devices	100+

Nur Faiza Ali *et al.*[2] proposed energy consumption is the most predominant problem to be analyzed in the sensor technology. Open flow protocol is the most important protocol for the SDN architecture. Open flow protocol enables the client and controller interaction of the forwarding plane effectively. Thus the flow of data packets from the switches to the controllers. The collection of information from the sensor nodes is obtained by the flow table information. Hence the data packets from sensor nodes can be effectively transferred from the routers to the controllers.

Wei *et al.*,[3] presented the energy efficient traffic problem in hybrid SDN/IP networks.[13] The algorithm which has IP routers performs the shortest path routing using a distributed routing table based on optimization the neighbor link and split the traffic at the SDN enabled switches by the overall controller.

Danda B. Rawat *et al.*,[4] proposed the architecture and security of SDN with the various strategies for the network implementation design. The performance of this SDN architecture from the standard network architecture is observed as shown in Figure 4. Another feature of SDN is openness for innovation open source platform which helps to improve further exploration, innovation, and development of the network. SDN is capable of upgrading the data center without losing the information, changing the overall workflow in the network and degrading overall performance.

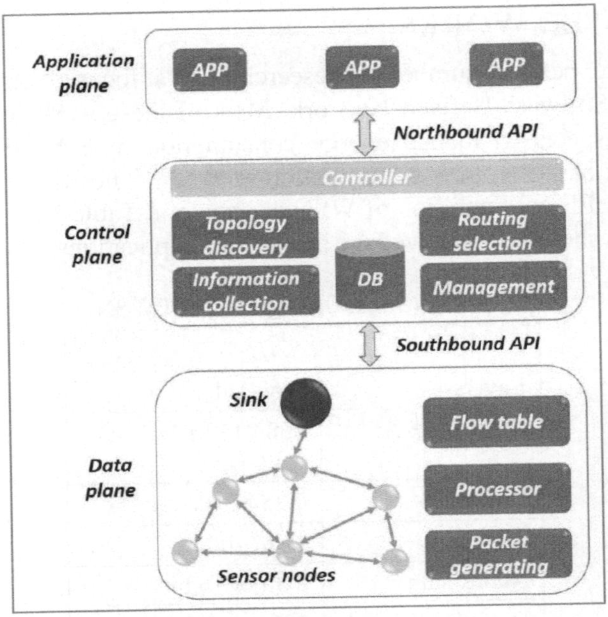

Fig. 4: SDN Layer Framework

The network design model of 25 nodes in Network Animator (NAM) file as shown in Figure 5, where node 0 is the source node, via intermediate node 3 and destination node 6 and the simulation parameters as shown in Table 2. The routing zone of these nodes is displayed in blue color and each node is displayed in a black color circle.

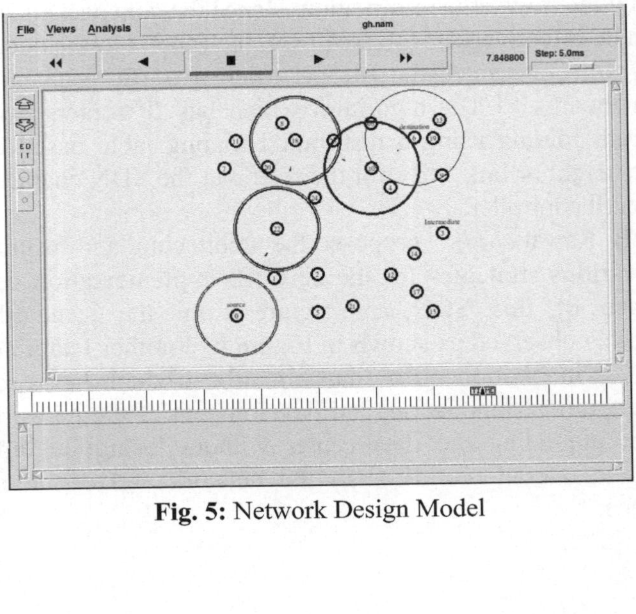

Fig. 5: Network Design Model

Table 2: Simulation Parameters

Type	*Parameter*	*Value*
Network	Area	100 × 100
	Location data sink	3
	Initial energy	2J
	N	25
Application	Data packet length	100 bits
	Broadcast packet length	25 bits

Xumin Huang,[6] a novel energy saving scheme that can flexibly control and route traffic relying on the difference of network device's energy-profile is discussed. The energy saving scheme[13] using OpenFlow makes use of the energy profile of network devices and switch port under various link rates.

Sukhveer Kaur *et al.*[7] The field of software defined networking is quite recent, but growing at a very fast pace. There are important research challenges to be addressed like security POX is a Python-based open source SDN Controller for developing SDN applications. POX controller can be used to run different applications like a switch, load balancer, and firewall. POX to demonstrate the dynamic programmability using a controller. If an organization requires a specific type of a network behavior it can develop or install application[8] to do what it needs, this application may be common networking function for example traffic engineering, policy routing, firewall, and security.

From the above-related works the following key observations are the existing energy optimization solutions do not consider the effect of routing algorithms on energy consumption calculation. In real time traffic scenario, at the time of route generation for any packet,[13] the controller may consider existing routing algorithms or create its own routing algorithm. For executing each routing algorithm some energy is required and that affects the energy consumption by a controller and thus by SDN.

These key observations motivate our present work, to develop a new energy optimization framework[9] for SDN that effectively addresses the above-mentioned problems without compromising the functionality of the network with varying requirement changes. In the next section, we present our proposed energy-efficient framework.

4. PROPOSED ENERGY-EFFICIENCY FRAMEWORK FOR SDN

In this section new energy efficient framework the key components of our framework are as follows: ERAS algorithm stands for energy efficient routing selection. This model has been observed as efficient for energy optimization.

4.1 ERAS Algorithm

Our proposed ERAS algorithm is two stacks of existing routing protocols[11] as input and returns the protocol with minimum link and controller energy.

- Stack1 consists of 2 Hierarchical routing protocols consist of Leach, Teen protocol.
- Stack2 consists of 2 Hybrid protocols consist of EIGRP, ZRP.

Hierarchical routing protocols are used for reduction of routing tables size which increases better scalability. Hybrid protocols find the best path for destination and consume less energy.

$$P_C = \beta * P_C^A + (1-\beta) * P_C^N \qquad \ldots (1)$$

$$P_C^A = P_{SA} + \sum_{i=1}^{n} W_i * CPUutil_I + f(T) * P_S \qquad \ldots (2)$$

$$P_C^N = P_{SA} + \sum_{i=1}^{n} W_i * CPUutil_I + P_{beacon} \qquad \ldots (3)$$

A Hierarchical protocol is used for the separation of a plane. This helps in analyses of routing table configuration. If the table matches with existing tables then the transfer of data packets is performed from a source node to the sink node. Leach is a clustering protocol[5] in which all nodes are periodically selected to be a cluster head and remaining as cluster members. The node with high power energy act as a Cluster head and all other becomes a cluster member.

Where,

P_C^A - Energy consumption by a controller when it wakes up itself

P_C^N - Energy consumption by a controller when it is activated by a neighbor.

P_C - Energy consumption by an ith controller

P_{beacon} - Energy of beacon message

$CPUutil_I$ - CPU utilization by controller due to different types of packets

The Teen protocol determines the threshold energy for the node which acts as a cluster head. Figure 6 shows that iterations with round number range from 0 to 100 for about 25 nodes have the average energy of each node is to be found to be around 0.099 and moderately reduced to a value of about 0.063 for more than 100 iterations. If minimization of energy is observed on a particular node below the threshold which remains as a cluster member. A Hybrid protocol is used to find the path for a destination.

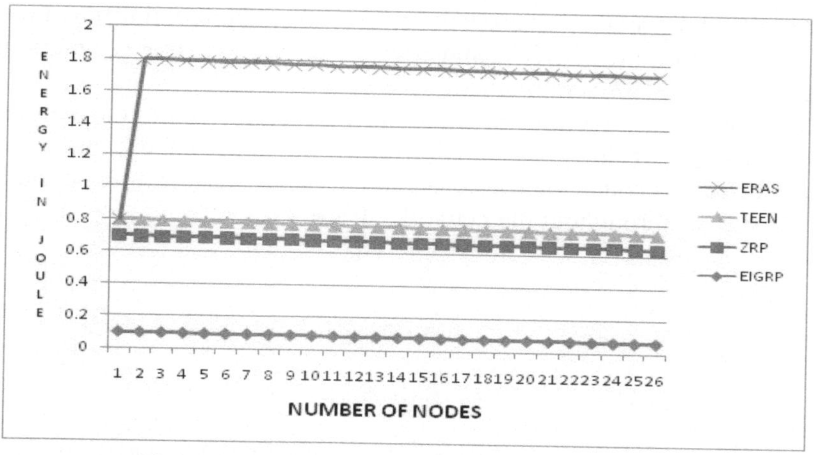

Fig. 6: Comparison with the Proposed Algorithm

ZRP protocol is used to transfer packets within the same zone using a routing table. Figure 6 shows the average energy of each node with round number ranges from 0 to 100 is 0.599 to 0.55 for about 25 nodes respectively. The routing table beyond the same zone is found to be difficult. CEDAR protocol is a routing protocol which maintains QoS throughout the network. CEDAR is a QoS aware protocol, core nodes are dominating the set of the network.

$$P_{SW}^{N} = P_{SA} + \sum_{i=1}^{n} P_{oi} + P_{beacon} \quad \ldots (4)$$

$$P_{SW}^{A} = P_{SA} + \sum_{i=1}^{n} P_{oi} + f(T) * P_{S} \quad \ldots (5)$$

$$P_{SW} = \alpha * P_{SW}^{A} + (1-\alpha) * P_{SW}^{N} \quad \ldots (6)$$

Where,

P_{SW}^{N} - Energy consumption by a switch when it is activated by a neighbor

P_{SW}^{A} - Energy consumption by a switch when it wakes up itself

P_{SW} - Energy consumption by the ith switch.

EIGRP protocol is a distance vector-based routing protocol. The average energy of each node round number ranges from 0 to 100 is about 0.099 to 0.015 efficiency, which is the same as the above iteration value for all 25 nodes. Thus the variation of EIGRP protocol is too small for more number of iterations. Due to vector algorithm throughput and packet loss can be reduced.

5. RESULTS AND DISCUSSION

We have implemented our proposed power consumption model and route selection algorithm in NS2 simulation platform with C^{++} programming language and OTCL scripting language.

From the observation of routing protocol, Figure 7 shows the algorithm with these protocols is compared with the iterative value of all the 25 nodes respectively and energy consumption ranges vary with these routing performances throughout the network.

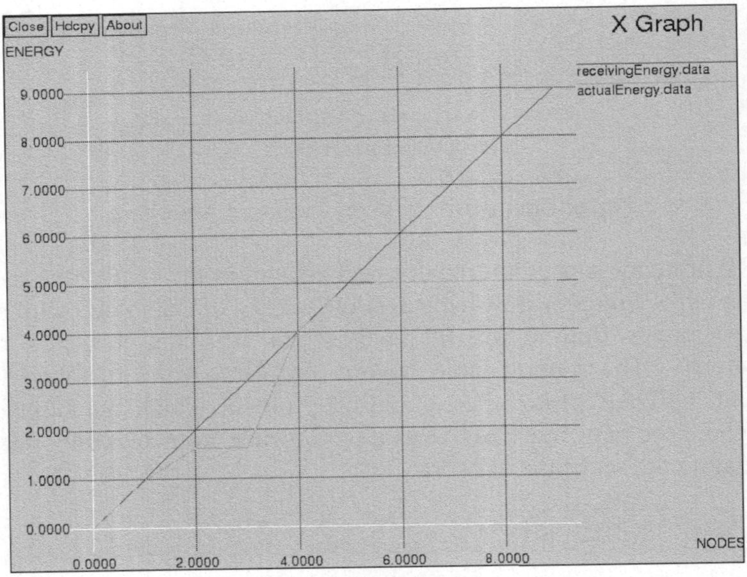

Fig. 7: XGraph Model

REFERENCES

[1] Kobo, Hlabishi I.; Abu-Mahfouz, Adnan M. and Hancke, Gerhard P., "A Survey on Software-Defined Wireless Sensor Networks: Challenges and Design Requirements", Published on February 8, 2017.

[2] Ali1, Nur Faiza; Said, Abas Md; Nisar, Kashif and Aziz, Izzatdin Abdul, "A Survey on Software Defined Network Approaches for Achieving Energy

Efficiency in Wireless Sensor Network", *2017 IEEE Conference on Wireless Sensors.*

[3] Xiang, Wei; Wang, Ning and Zhou, Yuan, "An Energy-Efficient Routing Algorithm for Software-Defined Wireless Sensor Networks", *IEEE Sensors Journal*, Vol. 16, No. 20, October 15, 2016.

[4] Rawat, Danda B. and Reddy, Swetha R., "Software Defined Networking Architecture, Security and Energy Efficiency: A Survey", *IEEE Communications Surveys & Tutorials,* Vol. 19, No. 1, First Quarter 2017.

[5] Priyadarsini, Madhukrishna; Bera, Padmalochan and Rahman, Mohammad Ashiqur, "A New Approach for Energy Efficiency in Software Defined Network", *2018 Fifth International Conference on Software Defined Systems.*

[6] Huang, Xumin; Yu, Rong; Kang, Jiawen; Xia, Zhuoquan and Zhang, Yan; Software Defined Networking for Energy Harvesting Internet of Things *IEEE Internet of Things Journal*, Vol. 5, No. 3, June 2018.

[7] Sukhveer Kaur , Japinder Singh and Navtej Singh Ghumman Network Programmability Using POX Controller.

[8] Hu, H., Chen, H.-H., Mueller, P., Hu, R.Q. and Rui, Y., "Software defined wireless networks (SDWN)": Part 1 [Guest Editorial]," *IEEE Communications Magazine*, Vol. 53, pp. 108–109, 2015.

[9] "SDN: Software Defined Networks Thomas D. Nadeau and Ken Gray August 2013 published by O'Reilly Media, Inc".

[10] Yick, J., Mukherjee, B. and Ghosal, D., Wireless sensor network survey, *Computer Network*, Vol. 52, No. 12, pp. 22922330, Aug. 2008.

[11] Khan, A.R., "Energy efficient protocol design issues in wireless sensor networks", *Journal of Information & Communication Technology,* Vol. 4, pp. 12–18, 2010.

[12] Chau Nguyen, Thi Minh; Hoang, Doan B. and Chaczko, Zenon, Can SDN technology be transported to software defined WSN/IoT? *2016 IEEE International Conference on Internet of Things (iThings) and IEEE Green Computing and Communications (GreenCom) and IEEE Smart Data (SmartData).*

[13] Wang, L. and Xiao, Y., "Energy saving mechanisms in sensor networks", in Broadband Networks, 2005. *BroadNets 2005. 2nd International Conference on*, 2005, pp. 724–732.

[14] Hu, H., Chen, H.-H., Mueller, P., Hu, R.Q. and Rui, Y., "Software-defined Wireless Networks (SDWN)", Part 1 [Guest Editorial], IEEE Communications Magazine, Vol. 53, pp. 108–109, 2015.

Effective Secondary User Selection Scheme for Underlay Cognitive Radio Network

R. Varshney, S. Vasuhi* and K. Dannesh

Department of Electronics Engineering, Madras Institute of Technology, Anna University, Chennai–600 044

*E-mail: vasuhi_s@annauniv.edu

ABSTRACT: *The recent advancement in wireless communication as created a greater problem of spectrum scarcity. To overcome the problem of spectrum scarcity, cognitive radio technology is a promising solution as it utilizes the spectrum in adaptable manner to increase the spectrum efficiency. Meanwhile, due to the broadcasting nature of cognitive radio network would poses security challenges to wireless communications. To enhance the security of the cognitive radio various scheme were introduced based on selecting the secondary user. The proposed scheme is based on the way the selected secondary user should ensure security. In underlay cognitive radio network, the primary user and the secondary user uses the spectrum simultaneously. Due the coexisting the secondary user should in turn protect the primary user from the eavesdropper. The selected secondary user must enable friendly jamming of the wiretap channel signal by including the interference. The interference added is in this case is going to beneficial for improving the secrecy rate of the cognitive radio network. The interference to transmitting signal is added in different ways and analyzed separately. The secrecy rate of the cognitive radio network has been analyzed under different user selection scheme.*

Keywords: Physical Layer Security, Cognitive Radio Networks, User Selection, Successive Interference Cancellation.

1. INTRODUCTION

Development of wireless technology has brought great demand towards the spectrum scarcity. The availability of radio frequencies are limited which is not enough for the developing wireless technology or the spectrum is idle most of the time. Hence to overcome the spectrum scarcity and problem of idle spectrum, a new technology has been developed called cognitive radio. In cognitive radio network, the spectrum can be efficiently used and also maintain the efficient communication in predominant condition.

Cognitive radio is possible to develop a radio that is able to look over the spectrum, detect which is freely available and implement the communication between the two end users. It is a smart radio transceiver hardware that automatically detects available channels in the spectrum, facilitates changing of the transmission and receiver parameters and allows the transmission in a given spectrum. This process is known as dynamic spectrum management. A cognitive radio may be defined as a radio that is aware of its environment, the internal state and with knowledge of these element any stored pre-defined objectives can make implement decisions about its behavior. It uses different technologies like adaptive and software defined radio.

Cognitive nature is broadcasting nature; it will be affected by various security attacks which in turn affects the performance of the network. As cognitive radio networks are open, dynamic and flexible in nature causes the system vulnerable to malicious attacks. Providing security to the network is not an easy task, this specifies various challenges and it is complicated and securing them is challenging one. The attacks may affect the functionality of the cognitive radio networks.

Physical layer is vulnerable to many threats. Although the three cognitive stages enable a cognitive user to learn its surrounding RF environment and adapt its transmission parameters to any changes in the environment, they are vulnerable to various attacks and introduce security threats. Eavesdropper is the attacker intercepts the confidential message between the users. The attack is vulnerable to transmission. Thus the attacker as the ability to tap and overhear the transmission between transmitter and receiver.

2. RELATED WORK

Several works have been reported by the researchers in order to provide security to the cognitive radio network. Xiaojun Tang *et al.*[1] proposed security model based on the property of the wireless medium, how the interference can encounter the eavesdropper and to provide security. The wiretap channel and the helping interferer (WT-HI) also considered. Broadcast and superposition are the fundamental properties of the wireless medium in which the broadcast nature makes the transmitted signal to be received by multiple receivers with different signal strength and the superposition property makes the receiver to observe a superposition of multiple simultaneous transmissions leading to interference. Using WT-HI model, the secrecy rate can be increased. As the secrecy rate depends on the interferer random encoding. Secrecy rate can be even increased when the main channel is worse than the wiretap channeled due to random code word. Only eavesdropper will only be interfered.

Zhaoyu Gao *et al.*[2] proposed model for the secondary user using the collaborative spectrum sensing method, there is possibility of location privacy threats due to Secondary User (SU) sensing about the Primary User (PU) is mainly based on physical location. Hence two different attacks has been evolved based on location namely RLC and DLC. To overcome these two attack two separate method has been introduced privacy preserving sensing report aggregation (PPSRA) against protocol RLC attack and distributed dummy report injection (DDRI) protocol against DLC attack. In PPSRA protocol, each secondary user performs its spectrum sensing among the spectrum and encrypts its sensing report with secret key and shares it to the Fusion Center (FC) after receiving the spectrum sensing report from the entire secondary user it will aggregate finally. This protocol address the RLC attacks since each sensing result is encrypted with the user's secret and the FC without having knowledge about individual user. In DDRI protocol, during the user leaving/joining phase, other users can use dummy sensing report which could be provided by the FC's own sensing result to replace their real sensing report. This protocol overcomes the problem of DLC attack. Hence using these two protocols RLC and DLC attack can be prevented with minimum overheads. These two protocols can individually use to overcome the physical location threats.

Yulong Zou *et al.*[3] proposed a system for securing the secondary user in the presence of the primary eavesdropper using relay selection scheme. There are two types of relay schemes namely single relay (SRS) scheme and multiple relay scheme (MRS). In single relay selection scheme, among the multiple single best relay is selected for providing security. In the multiple relay selection schemes multiple relays are enabled for simultaneous transmission. The SU uses the relay selection scheme to select the relay for the transmission in between the secondary transmitter and the receiver. The intended secondary receiver can decode the message from the relay but the eavesdropper cannot decode and the secrecy rate can be enhanced. As the spectrum sensing reliability increases secrecy rate also increases. This implies the trade-off between the security and the reliability of the secondary transmission in presence of the eavesdropping attack. MRS can achieve high secrecy rate compared to SRS

Hongjiang Lei *et al.*[4] modeled a system for providing security in underlay CRN. Simultaneous wireless information and power transfer technique is used to transmit information and to transmit power. To increase the security of the primary user here two scheme OAS and SAS is used. In OAS the channel state information is known. In SAS the channel state information is unknown. Secondary transmitter communicates using the energy harvested from the primary transmitter. Energy harvesting is also done. Secondary transmitter is equipped with the rechargeable battery. When the no. of antenna increases the performance of the system can be provided. Using this antenna selection scheme

Mahmoud Khasawneh *et al.*[5] proposed a model for cognitive radio network, where the SU can dynamically leave or join the network which leads to problem of security. As the SU leave network without any indication there is possibility for the malicious user to occupy the network as like SU. Hence in this paper, two level authentication schemes for communication are proposed. This scheme can mitigate attacks such as reflection attack, denial attack and man in middle attack. Among no of secondary users, SU is divided into clusters based on geographical locations and each clusters has cluster head. In each cluster, one node is chosen as cluster head and FC. The two level authentication done in FC and cluster head. The SU which wants to join network has to be authenticated before using the network. The joining node has to process authentication process at the two levels. Two methodologies are utilized to guarantee security which is public key infrastructure and symmetric key cryptography. The public key cryptography is utilized to ensure the security of the current node, this will continued until symmetric key is shared among the communication nodes. Each node encrypts and decrypts using same symmetric key. The security is provided using the two level authentication processes instead reduces the number of cryptographic process. It improves the secrecy rate where the authenticating and joining node accept key agreement.

3. METHODOLOGY

3.1 System Model

An underlay cognitive radio network is considered with pair of primary user and many secondary users.

Figure 1 illustrates the system model. Sharing licensed channel that is used by a pair of PU'S denoted by T and R, and a secondary user network consisting of N users and a receiver D. The primary transmitter T forwards the confidential message to the primary receiver R in the presence of eavesdropper E. all the users are equipped with the single antenna and operates in half duplex mode. The channel gain from T to R is represented as h_{TR} from T to D is represented as h_{TD} and from T to E is represented as h_{TE}. The entire channel experiences independent Rayleigh fading and assumes that the CSI of the primary and secondary link are available to secondary transmitter as discussed in[7,9].

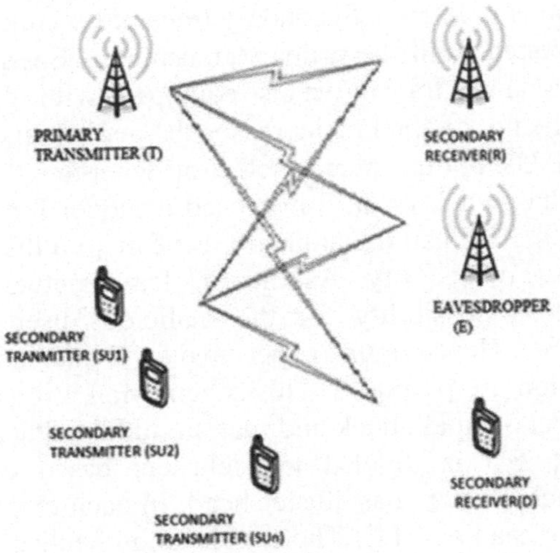

Fig. 1: Illustration of System Model

Ergodic secrecy rate (ESR) is used as the performance metric to improve the security of the cognitive radio network. The ESR of the network can be given as,

$$R_{S=} \{E[C_R] - E[C_E]\}^+ \quad \dots (1)$$

Where, R_S is the secrecy rate of the primary transmitter.
C_R is the achievable capacity of the primary receiver.
C_E is the achievable capacity of the eavesdropper.

3.1.1 *Minimal Interfernce Scheme*

The interference from the secondary user to the primary user must be below the threshold level. In minimal interference scheme the secondary transmitter interference will degrade the wiretap channel and also maintain the interference level below the threshold. The interference level at the eavesdropper is mainly due to the transmit power of the selected secondary user; the secondary user with worst channel is selected to communicate along with the primary user.

The transmit power of the selected secondary user is given by,

$$P_* = \begin{cases} P_S, |h_{S*R}| \leq \frac{I}{P_S} \\ \frac{I}{|h_{S*R}|}, |h_{S*R}| > \frac{I}{P_S} \end{cases} \quad \dots (2)$$

Select the secondary transmitter which can degrade the wiretap channel and also maintaining the interference to the primary user below the threshold level. The interference level at the eavesdropper is dependent on the selected secondary user transmit power.

The achievable secrecy rate is given using the Shannon capacity theorem. The achievable rate at the primary receiver, secondary receiver and eavesdropper is given respectively as,

$$C_R = \begin{cases} log\left(1 + \frac{\rho_T|h_{TR}|^2}{\rho_S|h_{S*R}|^2+I}\right), |h_{S*R}| \le \frac{\rho_I}{\rho_S} \\ log\left(1 + \frac{\rho_T|h_{TR}|^2}{\rho_I+I}\right), |h_{S*R}| > \frac{\rho_I}{\rho_S} \end{cases} \quad \ldots (3)$$

$$C_E = \begin{cases} log\left(1 + \frac{\rho_T|h_{TE}|^2}{\rho_S|h_{S*E}|^2+I}\right), |h_{S*R}| \le \frac{\rho_I}{\rho_S} \\ log\left(1 + \frac{\rho_T|h_{TE}|^2}{\frac{\rho_I|h_{S*E}|^2}{|h_{S*R}|^2}+I}\right), |h_{S*R}| > \frac{\rho_I}{\rho_S} \end{cases} \quad \ldots (4)$$

$$C_D = \begin{cases} log\left(1 + \frac{\rho_S|h_{S*D}|^2}{\rho_T|h_{TD}|^2+I}\right), |h_{S*R}| \le \frac{\rho_I}{\rho_S} \\ log\left(1 + \frac{\frac{\rho_I}{|h_{S*E}|^2}|h_{S*D}|^2}{\rho_T|h_{TD}|^2+I}\right), |h_{S*R}| > \frac{\rho_I}{\rho_S} \end{cases} \quad \ldots (5)$$

The secrecy rate of the primary user is analyzed under different cases.

Case 1: Achievable Ergodic Secrecy Rate:

To determine the achievable secrecy rate the secrecy rate computation is considered from.[11] The capacity of the primary user and secondary user is calculated from equation (3) & (4).

$$R_{S=} \{R_R - R_E\}^+ \quad \ldots (6)$$

$$R_{S=} \{E[C_R] - E[C_E]\}^+$$

Case 2: Impact of I on Secrecy Performance:

The achievable secrecy rate is determined on considering interference temperature. Assuming that the transmit power of the primary and secondary transmitter is larger enough. The achievable secrecy rate is determined using equation (6) from equation (3) & (4).

Case 3: Impact of P_S on Secrecy Performance:

When the interference temperature constraint is not considered then the selected secondary user power will be peak power. The achievable secrecy rate is determined using equation (6) from equation (3) & (4).

3.1.2 *Maximal Jamming Scheme*

The interference from the secondary user is not limited by interference temperature. The signal is transmitted with the peak power. Instead both the primary receiver and the eavesdropper is equipped with the Successive Interference Cancellation (SIC). The SIC is help in cancelling the interference from the transmitted signal. The channel state information is only known to primary and the secondary user but not in the eavesdropper. Hence eavesdropper cannot cancel the interference and provide security to the primary transmitter.

The secondary user with strongest interference channel is selected and the transmission rate the of selected secondary transmitter is given in the form of,

$$R_J = log\left(1 + \frac{\rho_T|h_{S\#R}|^2}{\rho_S|h_{TR}|^2+\sigma^2}\right) \quad \dots (7)$$

Where σ is the noise.

The capacity of the primary receiver and the eavesdropper are given respectively as,

$$C_R = log(1 + \rho_T|h_{TR}|^2) \quad \dots (8)$$

$$C_E = \begin{cases} log(\rho_T|h_{TE}|^2), \frac{|h_{S\#R}|^2}{|h_{TR}|^2} \le \frac{|h_{S\#E}|^2}{|h_{TE}|^2} \\ log\left(1+\beta\frac{|h_{TE}|^2}{|h_{S\#E}|^2}\right), \frac{|h_{S\#R}|^2}{|h_{TR}|^2} > \frac{|h_{S\#E}|^2}{|h_{TE}|^2} \end{cases} \quad \dots (9)$$

Where fixed ratio, $\beta = \frac{P_T}{P_S}$ and it hold $\beta > 1$.

When the secondary user with the strongest channel is considered, the achievable ESR of the primary user is approximated from equation (1).

3.1.3 *Mutual Jamming Scheme*

The security of both the primary and secondary user is ensured by using the secondary user has the relay node. Secondary user is considered as the trusted node. The primary transmits the signal to the secondary transmitter and secondary transmitter adds artificial noise to the original information and transmits to the receiver. Similarly way the secondary user transmits the signal to the secondary receiver. The power of the transmitter is increased and the security is provided to the cognitive radio network in this scheme.

(a) *Different Phases of Mutual Jamming Scheme:* Mutual jamming scheme operates in different phases. There are 3 different phase in this scheme:

Phase-1: Primary transmitter forwards its encoded message signal to secondary transmitter.

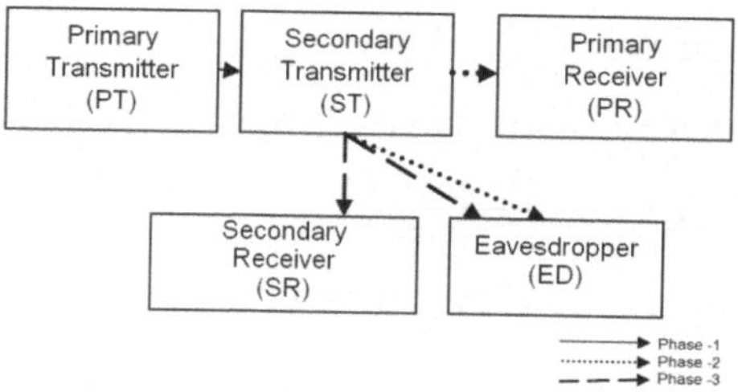

Fig. 2: Illustration of Phases of Mutual Jamming Scheme

Phase-2: Secondary transmitter on receiving the message signal from the primary transmitter, it adds the artificial noise to the original message signal and forwards to the primary receiver.

Phase-3: In this phase the secondary transmitter uses the spectrum for its own transmission. The secondary transmitter adds artificial noise to its own original message and forwards to the secondary receiver.

Phase 1: The primary transmitter decides to allocate time slot for its transmission from primary transmitter and secondary transmitter. The power of the primary transmitter is given by,

$$P_P = \frac{1}{N}\sum_{k=1}^{N} |S_k|^2 \leq P_{max} \qquad \text{... (10)}$$

Where the P_{max} is the maximum power of the primary transmitter, S is the primary message and confidential message is divided into an N-length block code word(S).

The information rate at the secondary transmitter from[12] and is given by,

$$R_{PS} = (1-\propto)log_2(1+\rho_{ps}) \qquad \text{... (11)}$$

Where $\propto$ is the time slot.

Phase 2: The primary transmitter decides to allocate time slot for its transmission from secondary transmitter and primary transmitter. The power of the primary transmitter is given by,

$$P_P = \frac{1}{n}\sum_{k=1}^{n} |S_k|^2 \leq P_{s,max}$$

Where the $P_{s,max}$ is the maximum power of the secondary transmitter, S is the primary message and confidential message is divided into an n-length block code word(S).

The information rate at the primary receiver is given by,

$$R_{SP} = (\propto \beta) log_2(1+\in \rho_{sp}) \quad \dots (12)$$

Where β is the time slot.

The information leakage at the eavesdropper is given by,

$$R_{SE} = (\propto \beta) log_2 \left(\frac{(1+\rho_{SE})}{(1+(1-\in)\rho_{se})}\right) \quad \dots (13)$$

The achievable primary secrecy rate is given by,

$$R_{PSEC} = R_{SP} - R_{SP} \quad \dots (14)$$

Phase 3: After removing the artificial noise at the secondary receiver. The information rate at the secondary transmitter is given by,

$$R_{SS} = (1-\propto) log_2(1 + \rho_{SS}) \quad \dots (15)$$

The leakage rate at the eavesdropper is given by,

$$R_{SE} = \propto (1 - \beta) log_2 \left(\frac{(1+\rho_{SE})}{(1+(1-\in)\rho_{se})}\right) \quad \dots (16)$$

The achievable primary secrecy rate is given by

$$R_{SSEC} = R_{SS} - R_{SE} \quad \dots (17)$$

4. RESULTS AND DISCUSSION

4.1 Minimal Interference Scheme

The performance of the minimal interference is studied and analyzed under different conditions.

4.1.1 *Achievable ESR of Minimal Interference Scheme*

Figure 3 shows that the secrecy rate of the primary user increases with no of user. This shows that the physical layer security using multiuser gain

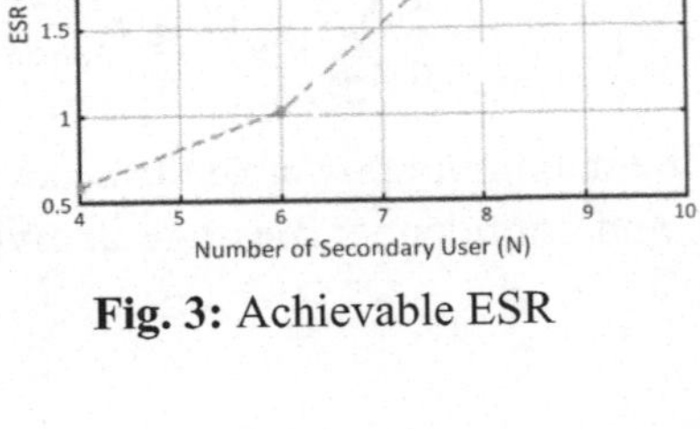

Fig. 3: Achievable ESR

can achieved. The secondary user selection improves secrecy performance of the users.

4.1.2 *Impact of I on ESR*

The relationship between the interference and the no of secondary user is analyzed. The secrecy performance is mainly based on the multiuser and the interference.

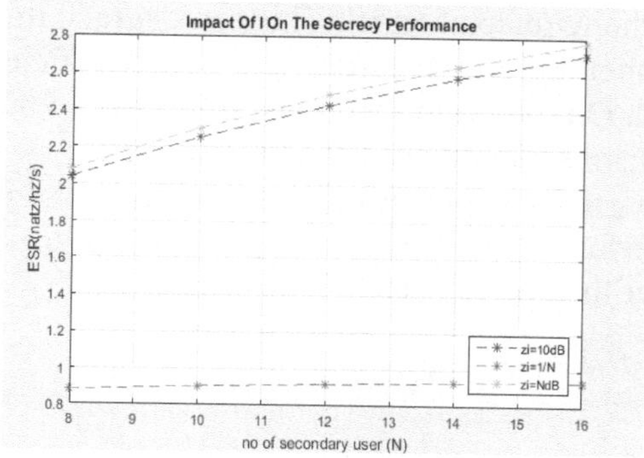

Fig. 4: Impact of I

Figure 4 shows that when the interference temperature and the no of secondary user are inversely proportional then performance is small. But when the interference temperature and the no. of secondary user is directly proportional then performance is large.

4.1.3 *Impact of* P_S *on ESR*

Even when the interference temperature is not limited by using multiuser in physical layer, the secrecy performance can be achieved. The multiuser gain can be achieved when the transmitting power is considered being small.

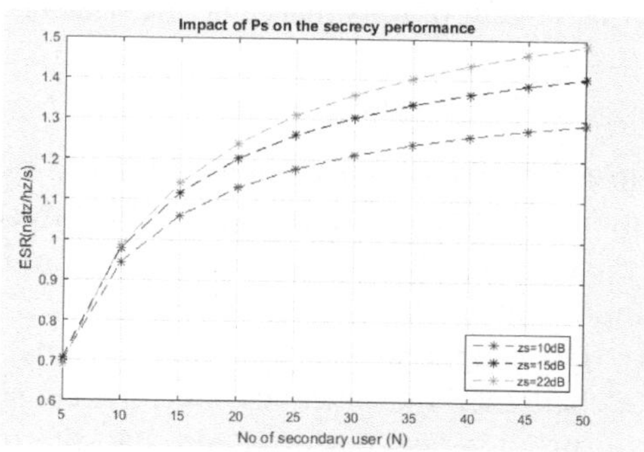

Fig. 5: Impact of P_s

Figure 5 shows that when the power of the secondary user increases with the increasing no of secondary user can achieve high secrecy rate. As increasing secondary user power implies the degradation of the wiretap channel.

4.2 Maximal Jamming Scheme

The power of the secondary user implies on the degradation of the performance of the wiretap channel. This can degrade the performance of the wiretap channel and the eavesdropper cannot efficiently overhear to the main channel. On increasing the transmit power of the secondary user increases the transmission rate of the secondary user.

Figure 6 shows that the power of the secondary user increases implies on the secrecy performance of the Primary user. And also by increasing no of secondary user the secrecy can also be increased.

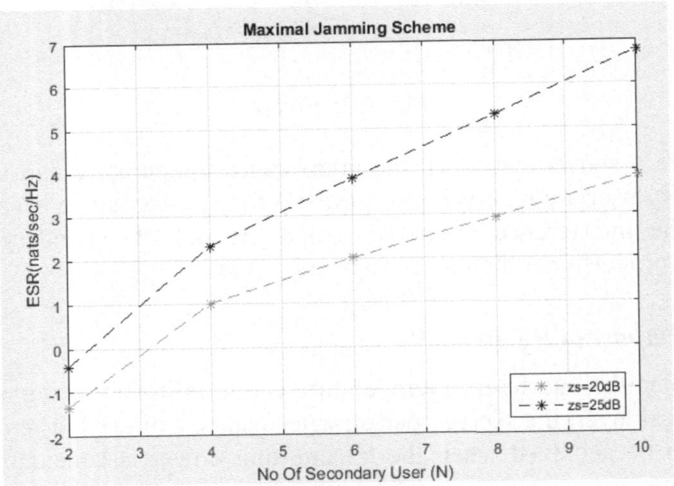

Fig. 6: ESR of Maximal Jamming Scheme

4.3 Comparison of Two Schemes

Both the scheme uses interference as the main factor to achieve the secrecy performance. But interference is kept below the threshold in minimal interference scheme and no limitation in maximal jamming scheme. Both the scheme is compared below.

Figure 7 shows the comparison between the minimal interference and maximal jamming scheme. In minimal interference scheme the interference control can provide less performance than the maximal jamming using no of secondary user.

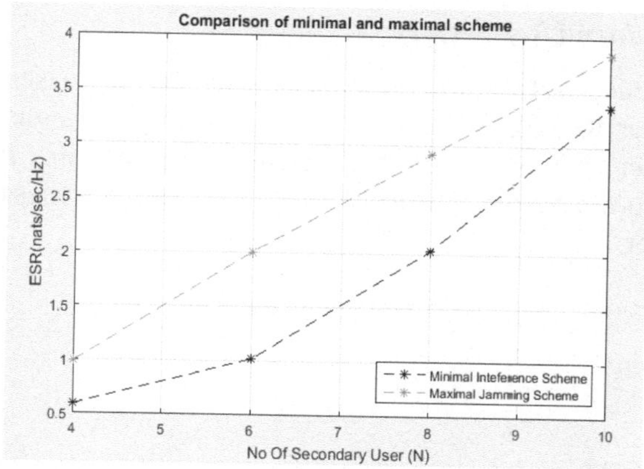

Fig. 7: Comparison of Minimal Interference and Maximal Jamming Scheme

4.4 Mutual Jamming Scheme

The transmit power of the secondary user can provide secrecy to both the user.

4.4.1 *Primary User Secrecy*

The power allocated to the secondary user by the primary user enables the secondary user to add artificial noise to the original signal so that the malicious user cannot obtain the original information.

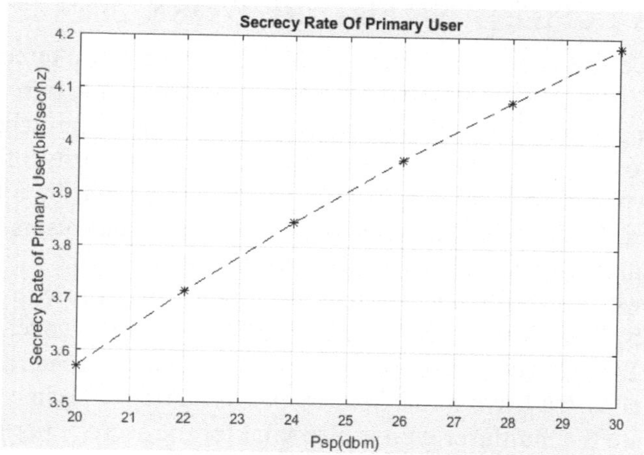

Fig. 8: Secrecy Rate of Primary User

Figure 8 shows that the power distributed to the secondary user from the primary user increases the secrecy performance can also be increased.

4.4.2 *Secondary User Secrecy*

The power allocated to the secondary user by increases enables the secondary user to add artificial noise to the original signal so that the malicious user cannot obtain the original information. The increased interference to the original information can prevent the message from the malicious user.

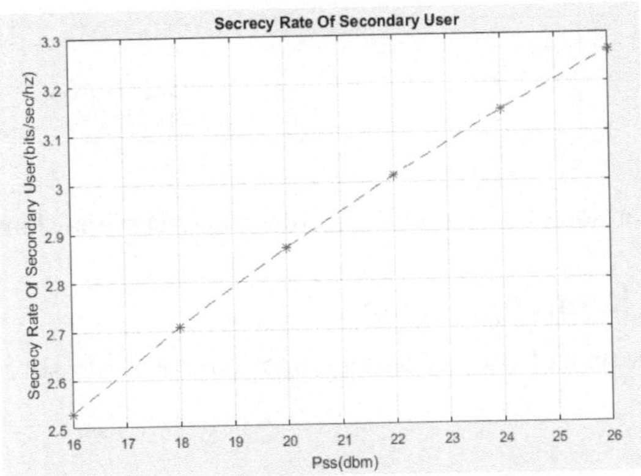

Fig. 9: Secrecy Rate of Secondary User

Figure 9 shows that the Power Allocated to the Secondary User Increases the Secrecy Performance can also be Increased

5. CONCLUSION AND FUTURE WORK

To enhance the security of the cognitive radio network three different secondary user selection schemes namely minimal interference based scheme, maximal jamming scheme and mutual jamming scheme have been proposed. The performance of physical layer security of cognitive radio network in the presence of both eavesdropper and the interference is analyzed. In the proposed system, selected secondary users maximize their transmission rate and provide security to both the users.

The power value of the transmitter also plays a major role in improving the secrecy rate. The secrecy rate at different power level has been analyzed. Experimental result demonstrates the impact of interference and peak power of the users in the secrecy performance. Experimental results also indicate the multiuser gain is also vital for the secrecy performance.

In future study, secure transmission with eavesdropper equipped with multiple antennas will be examined. That will enable us to study the secrecy performance of the cognitive network radio. Besides eavesdroppers there are some more threats to the physical layer security of CR networks such as PU emulation attack, objective function attack, learning attack, spectrum sensing data falsification attack. In future study, the research will extend to study of other physical layer threats.

REFERENCES

[1] Tang, X., Liu, R., Spasojevi¢, P. and Poor, H.V., "Interference Assisted Secret Communication", *IEEE Trans. Inf. Theory*, May 2011, Vol. 57, No. 5, pp. 3153–3167.

[2] Gao, Zhaoyu; Zhu, Haojin; Li, Shuai and Du, Suguo, Shanghai Jiao Tong University Xu Li, Inria Lille," Security and Privacy of Collaborative Spectrum Sensing in Cognitive Radio Networks", *IEEE Wireless Communications*, December 2012.

[3] Zou, Yulong, Benoit Champagne, Wei-Ping Zhu and Lajos Hanzo, "Relay-Selection Improves the Security-Reliability Trade-Off in Cognitive Radio Systems", *IEEE Transactions on Communications*, Vol. 63, No. 1, January 2015.

[4] Lei, Hongjiang; Xu, Ming; Ansari, Imran Shafique; Pan, Gaofeng; Qaraqe, Khalid A. and Alouini, Mohamed-Slim, "On Secure Underlay MIMO Cognitive Radio Networks with Energy Harvesting and Transmit Antenna Selection", *IEEE Transactions on Green Communications and Networking*, Vol. 1, No. 2, June 2017.

[5] Khasawneh, Mahmoud and Agarwal, Anjali, "A Secure and Efficient Authentication Mechanism Applied to Cognitive Radio Networks", *IEEE Access,* Vol. 5, 2017, August 22, 2017.

[6] Shu, Z., Qian, Y. and Ci, S., "On physical layer security for cognitive radio networks," *IEEE Netw.*, Vol. 27, No. 3, pp. 28_33, May/Jun. 2013.

[7] Mokari, N., Parsaeefard, S., Saeedi, H. and Azmi, P., "Cooperative securer source allocation in cognitive radio networks with guaranteed secrecy rate for primary users," *IEEE Trans. Wireless Commun.*, Vol. 13, No. 2, pp. 1058–1073, Feb. 2014.

[8] Zhang, N., Lu, N., Cheng, N., Mark, J.W. and Shen, X.S., "Cooperative spectrum access towards secure information transfer for CRNs," *IEEE J. Sel. Areas Commun.*, Vol. 31, No. 11, pp. 2453–2464, Nov. 2013.

[9] Chen, G., Gong, Y., Xiao, P. and Chambers, J.A., "Dual antenna selection in secure cognitive radio networks," *IEEE Trans. Veh. Technol.*, Vol. 65, No. 10, pp. 7993–8002, Oct. 2016.

[10] Deng, H., Wang, H.-M., Wang, W. and Lee, M.H., "Dual user selection for security enhancement in uplink multiuser systems," *IEEE Commun. Lett.*, Vol. 20, No. 9, pp. 1900–1903, Sep. 2016.

[11] Ban, T.W., Choi, W., Jung, B.C. and Sung, D.K., "Multi-user diversity in a spectrum sharing system," *IEEE Trans. Wireless Commun.*, Vol. 8, No. 1, pp. 102_106, Jan. 2009.

[12] Dong, L., Yousefi'zadeh, H. and Jafarkhani, H., "Cooperative jammingand power allocation for wireless relay networks in presence of eavesdropper," in *Proc. IEEE ICC*, Jun. 2011, pp. 1–5.

Supervisory Fuzzy Controller to Reduce the Energy of a Wastewater Treatment Plant

S. Saravana Kumar* **and K. Latha**
MIT, Chennai
*E-mail: sarvana.ssvm@gmail.com

ABSTRACT: *This paper presents a Fuzzy Logic Control (FLC) scheme for a biological waste water treatment plant, with the goal of reducing the effluent ammonia concentration and minimizing the aeration energy consumption. The famous Benchmark Simulation Modell (BSM1) is used to test the control strategy. PI controllers maintain the DO (Dissolved Oxygen) concentration in the aerobic tanks. The FLC controller provides set-point to DO controllers based on the ammonia concentration in the third tank and fourth tank. The proposed scheme has been tested under dry weather profile. To reduce the aeration energy consumption, two additional PI controllers are used in tank 3 and 4 respectively. Aeration energy of 180 kWh/d has been recovered from the aeration system. Substantial improvement in the reduction of aeration energy is resulted when compared with the benchmark PI controller. Simulation results show the benefits of the proposed method.*

Keywords: FLC, BSM1, ASP.

1. INTRODUCTION

The Waste Water Treatment Plant (WWTP) is aimed to remove the pollutants present in the water. In this paper, a WWTP for municipal waste water is considered. The common pollutants in the municipal waste water are organic materials, nitrogen compounds and suspended particles which when discharged to the environment causes serious damage. The natural way of treating the wastewater with the help of micro-organisms is called Activated Sludge Process (ASP). In ASP, the micro-organisms oxidize the organic materials to carbon dioxide and ammonia to nitrate. The WWTPs are complex nonlinear systems with varying time constants. The unpredictable behavior of the micro-organisms and the highly varying influent (flow and concentrations) make the control of WWTP a difficult task. In addition, the effluent standards must be kept within the limits as specified by the International Water Association (IWA).[11] There are various difficulties associated in the control of WWTP. The WWTP uses aerator to supply atmospheric air to the aerobic tank. This aeration system alone consumes about 50%–75% of the total plant energy.[21] The main objective of this work is to operate the WWTPs at a lower aeration energy profile.

The evaluation of various control techniques and its comparison is explained in the literature. The behavior of the WWTP is not constant due to the variability of influent, different time constant, complex biological phenomenon and lack of standard evaluation criteria. To overcome the above mentioned problems, a standard approach BSM1[11] was developed by International Water Association (IWA) Task Group on benchmarking of control strategies for WWTPs. Vilanova *et al.*[20] proposed a multilayered PI control strategy to achieve nitrogen removal at the satisfactory level. Yong *et al.*[18] developed six different control strategies for the integrated control of nitrate recirculation flow rate and external carbon flow in a pre-denitrifying tank. Yong *et al.* evaluated their control strategy using the European **Co**-operation in the field of **S**cientific and **T**echnical research (COST) simulation benchmark.[21] Meyer and *et al* developed a fuzzy control system to provide set point to the lower level PI controller.[23] Based on the literature, the standard PID controller alone is not effective to control the nonlinear WWTPs.

The complex nature in WWTP facilitated more advanced control strategies in order to improve the effluent quality, to optimize the use of energy resources and to follow the effluent discharge standards.[11] Wahab *et al.*[19] developed a multivariable PID controller to control DO in three aerobic tanks. Carlos *et al.*[2] also proposed an adaptive fuzzy control scheme to control DO in three aerobic tanks. Samsudin *et al.*[22] developed a non-linear PI control scheme to control the DO level in aeration tanks. With this control scheme, they achieved reasonable improvement in EQI. Only few literatures focused in keeping the effluent total nitrogen within the specified limit. Seocheon-dong *et al.*[6] developed a cascaded hierarchical control structure where a MPC was used to provide set-point to the lower level PI controller. This control scheme also aimed to keep the effluent total nitrogen within the limit. Also, Wenhao Shen *et al.*[24] proposed an ANFIS structure which keeps the effluent quality below the specified level. A multilayered control scheme developed by Vilanova *et al.*[20] focused in limiting the total nitrogen of the effluent. Other literature focused on varying the DO set-point[20, 4, 9] to improve the effluent standard and advanced control schemes like MPC[27, 7, 10] were proposed to improve the control performance of WWTP.

From the literature, it is inferred that few control schemes based on fuzzy was developed to keep the effluent quality under control. FLC is based on linguistic rules with an IF-THEN general configuration. FLC does not need a precise mathematical model of the nonlinear dynamic plant[3, 17] to be controlled. The FLC can be designed from the knowledge of the plant operator. So, FLC would be a promising choice for the non-linear plant control. This paper proposes the application of FLC in combination with ANFIS to reduce the aeration energy of the fifth reactor, thereby minimizing the overall energy requirement of the plant.

The paper is organized as follows. FLC is explained in Section 2. System configuration is described in Section 3 followed by the presentation of results and discussion in Section 4. The conclusions are drawn in Section 5.

2. CONTROL STRATEGY

2.1 Fuzzy Logic Control

Fuzzy logic theory is based on the strategy of expert knowledge to take decision which was first developed and proposed in.[15] FLC shown in Figure 1 is an interesting alternative to the conventional controllers such as PI.[3] A typical FLC consists of fuzzifier, rule base, inference engine and defuzzifier.[5]

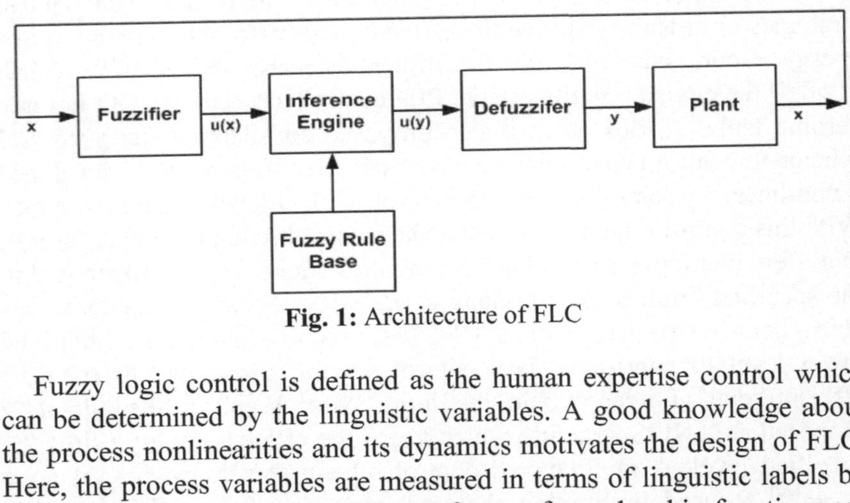

Fig. 1: Architecture of FLC

Fuzzy logic control is defined as the human expertise control which can be determined by the linguistic variables. A good knowledge about the process nonlinearities and its dynamics motivates the design of FLC. Here, the process variables are measured in terms of linguistic labels by membership functions. The different forms of membership functions are triangular, trapezoidal, Gaussian or the user defined function, chosen based on the experience. A triangular membership function is used here.

The fuzzy rule base or knowledge base is a set of if-then rules that store the experts' empirical knowledge, about the process operations. The fuzzification block computes the degree of membership of each input variable.

The grade of membership values denotes the strength of the variable in that fuzzy set (label). Based on the input variable, rules are triggered in the knowledge base. The inference engine aggregates the membership grades and the triggered rules using fuzzy operator. The defuzzifier converts the fuzzy set to a crisp value. The different defuzzification techniques available are centre of gravity, weighted average method, mean-max and so on. Here, the centroid method is used for defuzzification.

3. PROCESS DESCRIPTIONS

Wastewater Treatment Plants (WWTPs) are large non-linear systems subject to large perturbations in influent flow rate and pollutant load, together with uncertainties concerning the composition of the incoming wastewater. Currently BSM1 is accepted as the standard benchmarking tool.

3.1 Plant Layout

The benchmark is a simulation environment defining a plant layout, a simulation model, influent loads, test procedures and evaluation criteria. For each of these items, compromises were pursued to combine plainness with realism and accepted standards. The schematic of BSM1 is presented in Figure 2. The benchmark plant is composed of five activated sludge reactor with two anoxic tanks (1000 m^3) followed by three aerobic tanks (1333 m^3 each). The plant thus combines nitrification with pre-denitrification in a configuration that is commonly used for achieving biological nitrogen removal in full-scale plants. The activated sludge reactor is

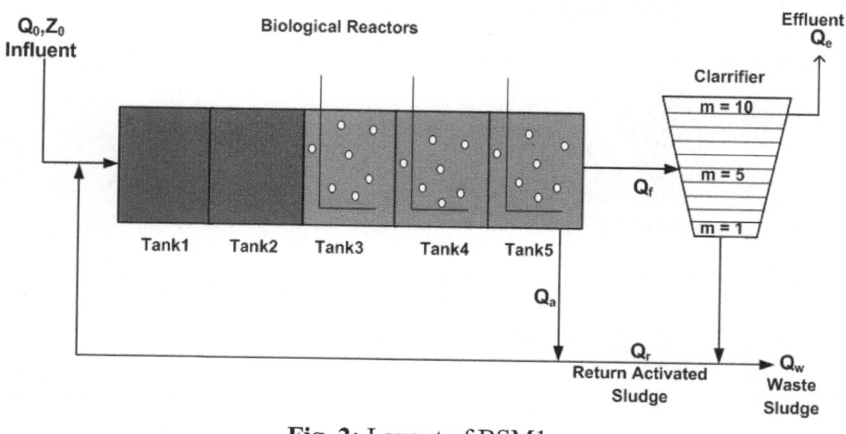

Fig. 2: Layout of BSM1

followed by a secondary settler (6000 m^3). The settler is modeled as ten layers with the height of 0.4 m and area of 1500 m^3. The 6th layer is the feed layer. The settler and reactors are modeled as double-exponential settling velocity function and activated sludge modell (ASM1)[2] respectively. No biological reaction is considered in the settler.

The plant is designed to operate for an average dry-weather influent flow rate of 18,446 m^3 with a hydraulic retention time of 14.4 hr. The waste sludge flow rate equals 385 m^3/day which corresponds to a biomass sludge age of 9 days. The ammonia removal is achieved through nitrification process which takes place in aerobic tanks. The denitrification process in anoxic tanks reduces the nitrate level in the effluent. A basic control strategy proposed in BSM1[11] is to test the benchmark: its aim is to control the dissolved oxygen level in the final compartment of the reactor by manipulation of the oxygen transfer coefficient and to control the nitrate level in the last anoxic tank by manipulation of the internal recycle flow rate.

3.2 Evaluation Criteria

A simulation protocol is established to ensure that results are obtained under the same conditions and can be compared. First, the plant stabilization has to be achieved in a closed loop for a period of 150 days, using constant influent data to drive the system to a steady state. Once the steady state is reached, a simulation with dry weather is run and finally the desired influent data (*i.e.* dry weather) is tested. Only the results of the last 7 days are considered for evaluation. The simulation procedure is described in Figure 4.

3.3 Energy Calculation

Mathematically, Aeration Energy (AE) is calculated as follows:

$$AE = \frac{S_O^{sat}}{1000T} \int_{t=7days}^{t=14days} \sum_{i=1}^{5} V_i * K_L a_i(t) dt \qquad \dots (1)$$

Where V_i is the volume of the reactor, S_O^{sat} is the maximum DO concentration that the waste water can hold and T is the total time.

Pumping Energy (PE) is given as,

$$PE = \frac{1}{T} \int_{t=7days}^{t=14days} \begin{pmatrix} 0.004 * Q_a(t) + 0.008 * \\ Q_r(t) + 0.05 * Q_w(t) \end{pmatrix} dt \qquad \dots (2)$$

Where $Q_a(t)$, $Q_r(t)$ and $Q_w(t)$ are internal circulation flow, activated sludge flow and waste sludge flow respectively.

4. SIMULATED RESULTS AND DISCUSSIONS

In this section, the proposed control scheme is tested and compared with the PI control scheme as specified in BSM1. Ideal sensors and actuators have been considered for the simulations. Mamdani method was selected for defuzzification. The fuzzy controller was designed based on the observation of the simulation results obtained by operating the plant in open loop condition and with the default control of BSM1. The plant is simulated as depicted in Figure 3. The proposed control scheme is shown in Figure 4. Three PI controllers control the DO concentration level in the aerobic tanks. In tank 5, the PI controller regulates DO at 2 mg (-COD)/l. The two PI controllers maintain the DO concentration in tank 3 and tank 4 respectively. Fuzzy system provides the set point to the above two DO PI controllers. The entire system is implemented and tested in Matlab Simulink (R2014a) environment.

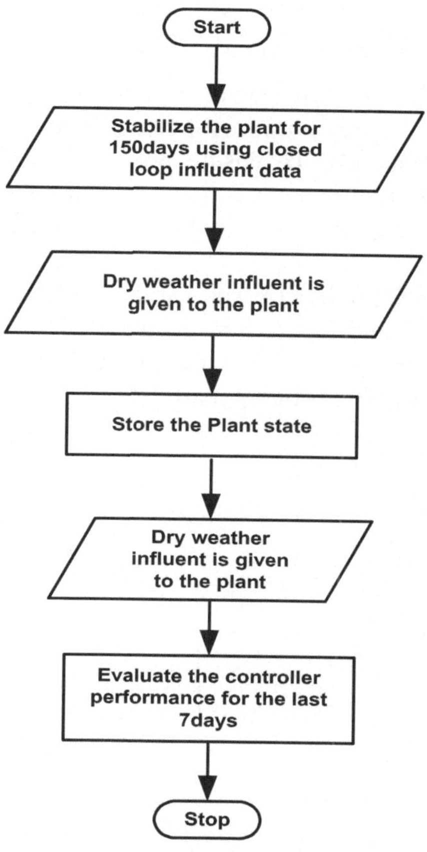

Fig. 3: Simulation Protocol

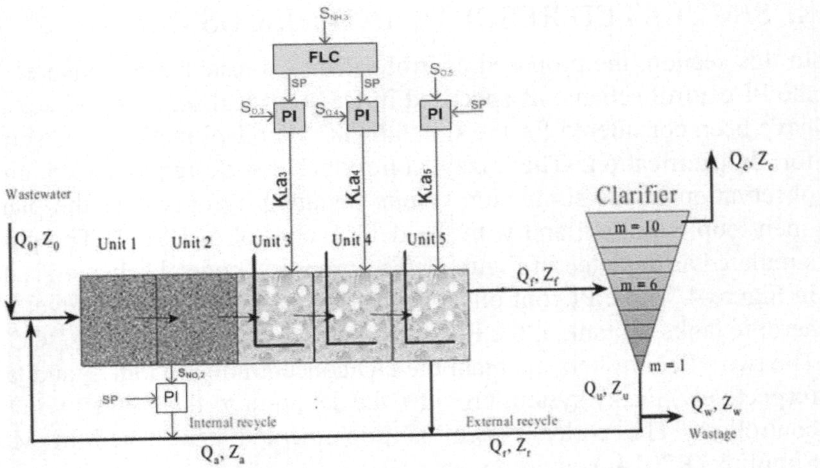

Fig. 4: Proposed Control Scheme

4.1 Ammonia Control in Aerobic Tanks

The PI controller in the fifth reactor regulates the DO concentration at 2 mg (-COD)/l as specified in BSM1 by manipulating the oxygen transfer coefficient (K_La_5). To ensure proper de-nitrification in anoxic tanks, DO concentration in tank 5 must be maintained in this level. PI controller in tank 3 an tank 4 manipulates K_La_3 and K_La_4 to maintain the DO concentration.

The manipulated variable K_La_5 and varying DO concentration of tank 5 is shown in Figure 5 and Figure 6 respectively. The PI controller effectively regulates the dissolved concentration at 2 mg (-COD)/l in tank 5 irrespective of the varying influent load.

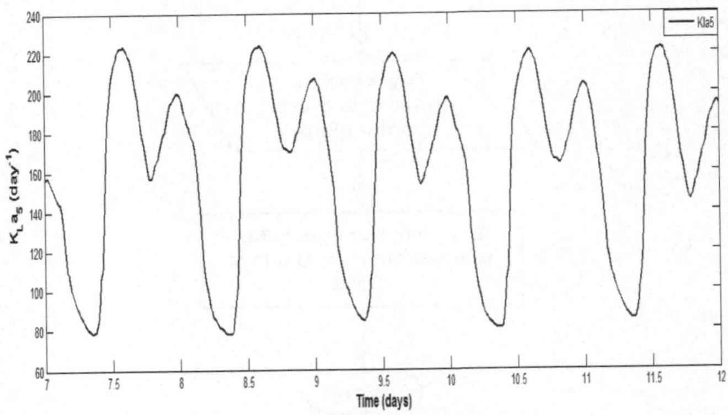

Fig. 5: K_La_5 Evolution for Dry Weather with PI Structure

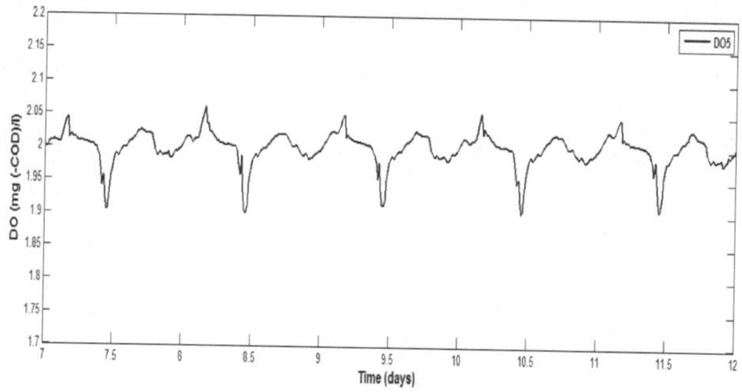

Fig. 6: DO Concentration of Tank 5 with PI Structure

FLC manipulates the set point of PI controllers for tank 3 and tank 4 based on the ammonia concentration in tank 3. FLC is designed in such a way that whenever the ammonia concentration ($S_{NH,3}$) in tank 3 increases, the set point level is also increased. Figure 7 shows the output of FLC for the varying ammonia concentration in tank 3. This will make sure that sufficient oxygen supply is supplied to oxidize SNH in tank 3 and tank 4. Figures 8 and 9 depict the set point tracking by the two PI controllers in tank 3 and tank 4 respectively. The low ammonia concentration in the plant lowers the set point levels to the two PI controllers. Thereby the aeration energy is minimized.

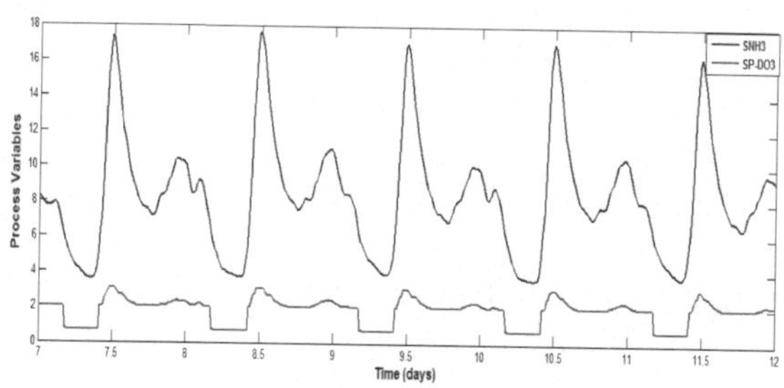

Fig. 7: Evolution of Set Point by FLC

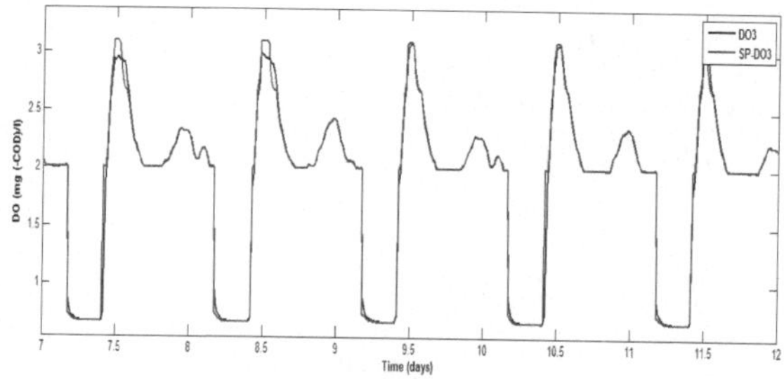

Fig. 8: Set Point Tracking of PI Controller in Tank 3

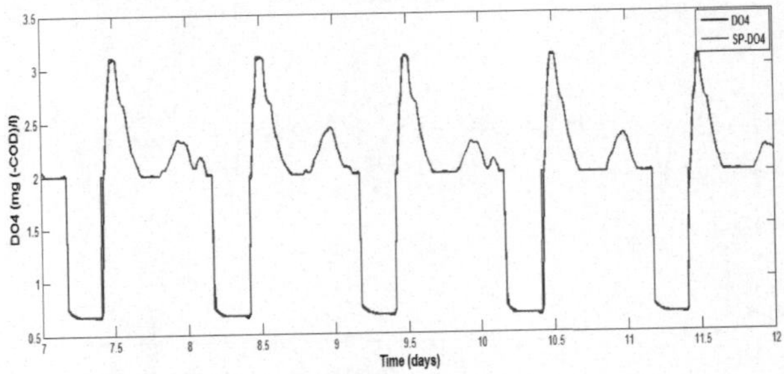

Fig. 9: Set Point Tracking of PI Controller in Tank 4

4.2 Nitrate Control

The nitrate concentration in tank 2 is regulated at 1 mg N/l by a PI controller. Figure 10 shows the manipulation of internal recirculation flow (Q_a) and Figure 11 depicts the nitrate concentration ($S_{NO,2}$) in tank 2.

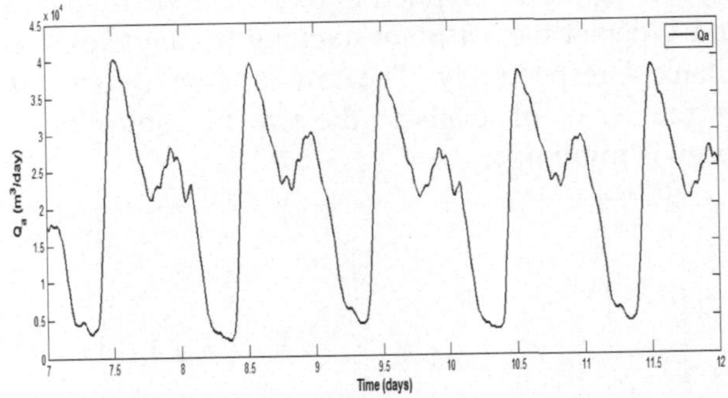

Fig. 10: Internal Recirculation Flow, Q_a in Tank 2

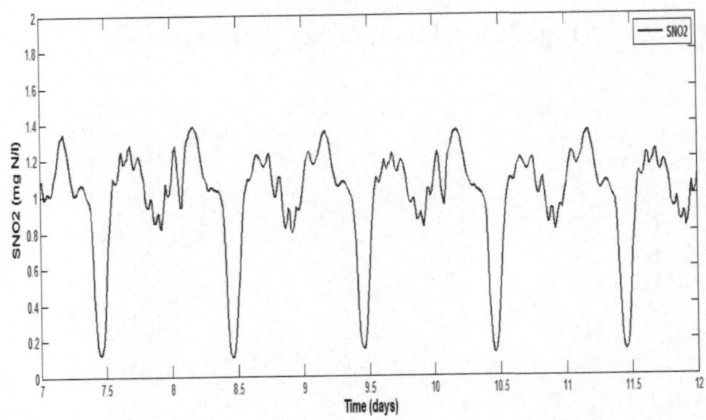

Fig. 11: Nitrate Concentration, $S_{NO,2}$ in Tank 2

The PI controller could not regulate nitrate concentration at 1 mg N/l satisfactorily but it performed better than the PI configuration mentioned in BSM1.

4.3 Energy Analysis

Energy analysis for both FLC and PI control schemes has been done. Table I shows the energy requirements of the proposed scheme and PI control scheme in the dry weather condition. Energy requirement of the aeration system in the proposed scheme is 180 kWh/d less than that of its counterpart. With the proposed scheme, the aeration energy is reduced by 4.86% in dry weather which could save more money for an entire year. The pumping energy in the proposed scheme is 2 kWh/d more than the one mentioned in BSM1. In spite of the small increase in pumping energy, the proposed scheme reduced aeration energy to a greater extent.

Table 1: Energy Analysis

	PI	*Proposed*
AE (kWh/d)	3697.7	3517.5
PE (kWh/d)	241.92	243.2

5. CONCLUSION

In this paper, a FLC-PI control scheme was presented to reduce the aeration energy of the plant. FLC was designed by embedding this control objective in its knowledge base. The performance of the proposed controller was compared and evaluated with the benchmark PI control scheme. The proposed scheme was simulated and tested under dry weather disturbances. The proposed controller saved approximately 180kWh energy per day from the aeration system. In spite of the decrease in aeration energy, the proposed scheme resulted in a better effluent quality than the one in BSM1. Thus, the proposed scheme provided best results in terms of aeration energy consumption.

The above work can be extended by formulating the control objectives can as an objective function. The minimization/maximization of the objective function can be carried out by manipulating the rules of FLC. To support the optimization algorithm, supervised learning algorithm can be designed to acquire the process knowledge. The future work may improve the effluent quality in a better way with minimal energy resources.

REFERENCES

[1] Abdulshahed, Ali M.; Longstaff, Andrew P. and Fletcher, Simon, "The application of ANFIS prediction models for thermal error compensation on CNC machine tools", *Applied Soft Computing*, Vol. 27, pp. 158–168, Nov. 2014.

[2] Coelho Belchior, Carlos Alberto; Matos Araujo, Rui Alexandre and Cardoso Landeck, Jorge Afonso, "Dissolved oxygen control of the activated sludge wastewater treatment process using stable adaptive fuzzy control", *Computer and Chemical Engineering*, Elsevier, pp. 152–162, Oct. 2011.

[3] Mamdani, E.H., "Application of fuzzy algorithms for control of simple dynamic plant", *Proceedings of IEEE*, Vol. 121, No. 12, pp. 1585–1588, 1974.

[4] Hernandez-del-Olmo, Felix; Gaudioso, Elena and Nevado, Antonio, "Autonomous Adaptive and Active Tuning Up of the Dissolved Oxygen Setpoint in a Wastewater Treatment Plant Using Reinforcement Learning", *IEEE Transactions on Systems, Man, and Cybernetics - Part C: Applications and Reviews*, Vol. 42, No. 5, pp. 768–774, Sep. 2012.

[5] Chen, G. and Pham, T., "Introduction to Fuzzy Sets, Fuzzy Logic, and Fuzzy Control Systems", *CRC Press*, Nov. 2000.

[6] Liu, Hongbin and Yoo, ChangKyoo, "Performance assessment of cascade controllers for nitrate control in a wastewater treatment process", *Korean Journal of Chemical Engineering*, pp. 667–673, 2011.

[7] Han, Hong-Gui; Qian, Hu-Hai and Qiao, Jun-Fei, "Nonlinear multiobjective model-predictive control scheme for wastewater treatment process", *Journal of Process Control*, pp. 47–59, Jan. 2014.

[8] Vanhooren, H. and Nguyen, K., "Development of a simulation protocol for evaluation of respirometry-based control strategies", Technical Report, University of Gent, Belgium, 1996.

[9] Santin, I., Pedret, C. and Vilanova, R., "Applying variable dissolved oxygen set point in a two level hierarchical control structure to a wastewater treatment process", *Journal of Process Control*, pp. 40–55, Mar. 2015.

[10] Santin, I., Pedret, C. and Vilanova, R., "Fuzzy Control and Model Predictive Control Configurations for Effluent Violations Removal in Wastewater Treatment Plants", *Industrial and Engineering Chemistry Research*, pp. 2763–2775, Feb. 2015.

[11] Alex, J., Benedetti, L., Copp, J., Gernaey, K.V., Jeppsson, U., Nopens, I., Pons, N., Rieger, L., Rosen, C., Steyer, J.P., Vanrolleghem, P. and Winkler, S., "Benchmark Simulation Model No. 1 (BSM1)", Technical Report, Department of Industrial Electrical Engineering and Automation, Lund University, Apr. 2008.

[12] Copp, J.B., "The Cost Simulation Benchmark: Description and Simulator Manual (COST Action 624 and Action 682)", *Journal Office for Official Publications of the European Union: Luxembourg,* pp. 1–154, 2002.

[13] Copp, J.B., "The Development of standardised influent files for the evaluation of activated sludge control strategies", *IAWQ Scientific and Technical Report*, 1999.

[14] Jang, J.S.R., "ANFIS: adaptive-network-based fuzzy inference system", *IEEE Transactions on Systems, Man and Cybernetics*, Vol. 23, No. 3, pp. 665–685, Jun. 1993.

[15] Zadeh, L.A., "Fuzzy sets, Information and Control", Vol. 8, No. 3, pp. 338–353, 1965.

[16] Henze, Mogens; Gujer, Willi; Mino, Takashi and Loosdrecht, Mark van, "Activated Sludge Models ASM1, ASM2, ASM2d and ASM3", 2002.

[17] Fuente, M.J., Robles, C., Casado, O., Syafiie, S. and Tadeo, F., "Fuzzy control of a neutralization process", *Engineering Applications of Artificial Intelligence*, Vol. 16, pp. 905–914, Mar. 2006.

[18] Yong, M., Yongzhena, P., Jeppsson, U., "Dynamic evaluation of integrated control strategies for enhanced nitrogen removal in activated sludge processes", *Control Engineering Practice*, Elsevier, pp. 1269–1278, 2006.

[19] Wahab, N.A.; Katebi, Reza and Balderud, Jonas, "Multivariable PID control design for activated sludge process with nitrification and denitrification", *Biochemical Engineering Jounal*, pp. 239–248, 2009.

[20] Vilanova, Ramon; Katebi, Reza and Wahab, NoraLiza, "N-Removal on Wastewater Treatment Plants: A Process Control Approach", *Journal of Water Resource and Protection*, Vol. 3, pp. 1–11, Jan. 2011.

[21] Ryan Hamilton, Bill Braun, Rondare, Ben Koopman, S. Spyrosa, Voronos, "Control issues and challenges in Wastewater Treatment Plants", *IEEE Control System Magazine*, pp. 63-69, Jul. 2006.

[22] Samsudin, S.I., Rahmat, M.F. and Wahab, Norhaliza Abdul, "Nonlinear PI Control with Adaptive Interaction Algorithm for Multivariable Wastewater Treatment Process", *Journal of Mathematical Problems in Engineering*, Hindawi publications, pp. 1–14, 2014.

[23] Meyer, U. and Popel, H.J., "Fuzzy control for improved nitrogen removal and energy saving in WWTPs with pre-denitrification", *Water Science Technology*, Vol. 47, No. 11, pp. 69–76, Feb. 2003.

[24] Shen, Wenhao; Tao, Erpan; Chen, Xiaoquan; Liu, Dawei and Liu, Hongbin, "Nitrate control strategies in an activated sludge wastewater treatment process", *Korean Journal of Chemical Engineering*, pp. 386–392, 2014.

[25] Shen, Wenhao; Chen, Xiaoquan and Corriou, Jean Pierre, "Application of model predictive control to the BSM1 benchmark of the wastewater treatment process", *Computer and Chemical Engineering*, Elsevier, pp. 2849–2856, Mar. 2008.

Author Index

9 789389 934199

Printed at Repro India Ltd.